# SpringerBriefs in Energy
## Energy Analysis

**Series Editor:** Charles A.S. Hall

For further volumes:
http://www.springer.com/series/10041

# Spain's Photovoltaic Revolution

## The Energy Return on Investment

by
Pedro A. Prieto and Charles A.S. Hall
with the assistance of Rigoberto Melgar

 Springer

Pedro A. Prieto
Vice President of the Asociación para el
    Estudio de los Recursos Energéticos
    (AEREN)
Member of the Board of ASPO
    International
Pozuelo de Alarcón
28223 Madrid, Spain

Charles A.S. Hall
Faculty of Environmental and Forest
    Biology and Graduate Program in
    Environmental Science
College of Environmental Science and
    Forestry
State University of New York
Syracuse, NY 13210, USA

ISSN 2191-5520           ISSN 2191-5539 (electronic)
ISBN 978-1-4419-9436-3   ISBN 978-1-4419-9437-0 (eBook)
DOI 10.1007/978-1-4419-9437-0
Springer New York Heidelberg Dordrecht London

Library of Congress Control Number: 2012951424

Printed on acid-free paper

Springer is part of Springer Science+Business Media (www.springer.com)

# Preface

We all know that the fossil fuels that sustain modern human civilization are finite and polluting. We also know that the amount of sunshine that enters the Earth surface is immense, many orders of magnitude greater than human needs. These two facts lead to one obvious conclusion: we need to replace fossil fuels with solar energy. There are many solar advocates who see the issue as essentially a "no brainer."

But others understand that solar is far from cheap in monetary terms. The first author, with a great deal of experience in the solar industry, understands this especially well, as he had to sign every purchase order as chief engineer for a series of major solar installations in Spain. In this book, we find that the best way to understand the fossil fuel subsidies that underlie whatever we do in our economy is to "follow the money." Anywhere in society that money is spent, energy (and mostly this means fossil fuels today) must be used to generate the goods and services that the money represents. In general, we consider money to be a "lien on energy" where general economic agreements allow the bearer of money to purchase energy-requiring goods and services. Manufacturers will purchase and use energy to put goods and services on the shelf in anticipation of sales. Take out the money, and an economy is still possible through barter. Take out the energy, and the economy stops, as the Cubans found in 1988 when the Russians cut off their oil and food disappeared from the stores within a week.

In this book, we attempt to evaluate all (or as many inputs as is possible) of the energy that goes into providing electricity for real-life photovoltaic systems in Spain, the country with the second largest installations of solar power as of 2008. Many of the inputs that we include are normally skipped in typical life cycle analyses (LCE) and energy payback time (EPBT) studies that have been made of solar PV power. While we believe this larger view of inputs to make PV is essential for a comprehensive energy analysis, we recognize that some of our inputs will be controversial. We leave it to the reader and to future analysts to make their own decisions about inclusivity and methods in general for a comprehensive analysis of EROI. Whatever your opinion, this study should really open your eyes to the degree to which fossil fuels underlie everything we do in our technological society.

A series of excellent photographs that help one to get a feel for photovoltaic power in Spain can be found at http://www.flickr.com/photos/87892847@N03/

# Acknowledgments

We thank Carlos de Castro, professor of Applied Physics and Energy Group, and System Dynamics Modeling Energy Group, University of Valladolid, for reviewing the manuscript and other valuable input.

Antonio Ruiz de Elvira, Professor of Applied Physics, University of Alcala de Henares, For peer reviewing and encouraging this venture and pioneering the EROI works in Spain.

Jim MacInnes provided an indepth revision and many helpful comments on transformity.

We also thank Ajay Gupta for reading and commenting on the manuscript.

# About the Authors

**Pedro A. Prieto** is a telecom technical engineer. He has worked in the telecom sector for 30 years as Development Engineer and Professor in the ITT R&D Labs, Commercial Director, Export Director, and Vice President of the radio communication division in Alcatel.

Pedro is Vice President of a not-for-profit organization, the Asociación para el Estudio de los Recursos Energéticos (AEREN), whose media is the web page Crisis Energética (www.crisisenergetica.org), which is working as an open space for debate and communications on energy issues and their role in demography, development, economy, and ecology and is a reference in the Spanish speaking world for ASPO (the Association for the Study of Peak Oil and Gas, an association of scientists of all over the world, devoted to the analysis and study of the global peak oil and gas and its consequences for mankind). AEREN represents ASPO in Spain. He attends many conferences on peak oil and renewable energies.

Pedro is a board member at ASPO International (www.peakoil.net) and is also member of the not-for-profit organization Científicos por el Medio Ambiente (CiMA) (www.cima.es.org).

Since 2006, Pedro A. Prieto has led several solar photovoltaic projects in Spain, a leading world country in solar PV penetration, covering today about 3 % of the total national electric demand, with about 4,246 MW installed base, of which he partially owns and completely manages a 1 MW modern feed-in plant in operation. He has assessed public and private entities in several projects for more than 30 MW, and has supervised, offered, and dimensioned many other similar solar photovoltaic projects.

In 2009, he was engaged as director of development for alternative energies at a listed Spanish Corporation devoted to the fields of telecommunications, technology, infrastructures, and renewable energies.

At present, Pedro works as a truly independent consultant in solar PV projects. He would like to dedicate this book to his granddaughters, Sophie and Chloe, whose smiles greatly enlighten the challenging times to come.

**Charles A.S. Hall** is professor of environmental science at the State University of New York College of Environmental Science and Forestry in Syracuse. He received his Ph.D from the University of North Carolina at Chapel Hill under Dr. H.T. Odum. His fields of interest are systems ecology, energy, and biophysical economics. Dr. Hall is author, coauthor, or editor of 10 books and 270 scholarly articles that have appeared in *Science, Nature, Bioscience, American Scientist*, and other scientific journals. He is best known for his development of the concept of EROI, or energy return on investment, which is an examination of how organisms, including humans, invest energy into obtaining additional energy to improve biotic or social fitness. He has applied these approaches to carbon balance, fish migrations, tropical land use change, and the extraction of petroleum and other fuels in both natural and human-dominated ecosystems. Presently, he is developing a new field, biophysical economics, as a supplement or alternative to conventional neoclassical economics while applying systems and EROI thinking to a broad series of resource and economic issues.

# Contents

# Table of Equivalences

| Kilo (K) | Mega (M) | Giga (G) | Tera (T) | Peta (P) | Exa (E) |
|---|---|---|---|---|---|
| $10^3$ | $10^6$ | $10^9$ | $10^{12}$ | $10^{15}$ | $10^{18}$ |

*Units*

1 metric tonne = 2204.62 lb. = 1.1023 short tons

1 kilolitre = 6.2898 barrels

1 kilolitre = 1 m$^3$

1 kilocalorie (kcal) = 4.187 kJ = 3.968 Btu

1 kilojoule (kJ) = 0.239 kcal = 0.948 Btu

1 British thermal unit (Btu) = 0.252 kcal = 1.055 kJ

1 kilowatt-hour (kWh) = 860 kcal = 3600 kJ or 3.6 MJ = 3,412 Btu

*Calorific equivalents*

One tonne of oil equivalent (toe) equals approximately:

| | |
|---|---|
| Heat units | 10 million kcal |
| | 42 GJ |
| | 40 million Btu |
| Solid fuels | 1.5 tonnes of hard coal |
| | 3 tonnes of lignite |
| Gaseous fuels | See natural gas and LNG table |
| Electricity | 12 MWh |

One million tonnes of oil produces about 4,400 GWh (= 4.4 terawatt hours) of electricity in a modern power station

# Abbreviations and Acronyms

AEF       Asociación Empresarial Fotovoltaica. It is one of the two largest Spanish associations of the photovoltaic sector (http://www.aefotovoltaica.com/).

a-Si      Amorphous silicon, a technique of PV cells for solar modules.

ASIF     Asociación de la Industria Fotovoltaica. It is one of the two largest Spanish associations of the photovoltaic sector (http://asif.org/).

BAU     Business as usual. The normal execution of operations within an organization.

BoS      Balance of System. It encompasses all components of a photovoltaic system other than the photovoltaic panels. This includes wiring, switches, support racks, an inverter, and batteries in the case of off-grid systems. In the case of free-standing systems, land is sometimes included as part of the BOS (*Source*: Wikipedia).

CNE     Comisión Nacional de Energía (www.cne.es). The regulatory body of energy systems in Spain.

CPI      Consumer Price Index.

CSP     Concentrated solar power. Plants generating electric energy by means of concentrating the sunrays, either on a central tower by means of two-axis tracking mirror fields or by means of parabolic trough mirrors on an axis transporting a fluid, generally a synthetic oil that goes into a molten salt deposit, acting as an energy buffer, which then exchanges heat into steam to power a turbine.

CUR     Comercializadoras de Último Recurso. Last resource commercial companies. They are legal, commercial entities—usually spin-offs of the main electric operators, doing the forecast production on behalf of those solar PV or wind producers that have not chosen a specific one.

EJ       Exajoule. A measure of energy equivalent to $10^{18}$ J. For reference, 1 J is 1 W/s.

Eout     Energy output or the energy that a system delivers in a given period of time.

| | |
|---|---|
| EPBT | Energy payback time. The time needed to recover the energy originally invested to generate a given amount of energy from an energy source. |
| EPC | Engineering, Procurement, and Construction companies. |
| ER | Energy return or energy returned. |
| EROEI | Energy return on the energy invested or sometimes called energy return on investment (*EROI*). It is the ratio of the amount of usable energy obtained from a particular energy resource to the amount of energy invested to obtain that energy. The former refers, exclusively, to energy ratios, while the latter refers also to energy return on economic ratios. |
| EROI | See *EROEI*. |
| GDP | Gross domestic product (GDP) refers to the market value of all final goods and services produced in a country in a given period. |
| GSM | Global system for mobile. It is a standard set developed by the European Telecommunications Standards Institute (ETSI) to describe protocols for second generation (2G) digital cellular networks used by mobile phones (*Source*: Wikipedia). |
| GPRS | General packet radio service. A packet-oriented mobile data service. |
| GW | A measure of power. It is a billion W. Its related energy level is giga-watt $\times$ hour (GWh). |
| HCPV | High-concentration photovoltaic systems. Systems using photovoltaic cells by concentrating the sunrays on them, generally with Fresnel lenses, to produce more energy with the same cell. They vary from low concentration (3 suns) to very high concentration (1,000 suns). |
| IEA | International Energy Agency. The OECD energy body. |
| IGBT | High-voltage insulated gate bipolar transistor. |
| INE | Instituto Nacional de Estadística. The Spanish public entity for national and regional statistics (http://www.ine.es/). |
| IP | Internet protocol. |
| LCA | Life cycle assessment. It is a procedure to assess environmental impacts associated with all the stages of a product's life from cradle to grave (i.e., from raw material extraction through materials processing, manufacture, distribution, use, repair and maintenance, and disposal or recycling). |
| mc-Si | Multi-crystalline silicon, also known as polycrystalline silicon, a technique of PV cells for solar modules. |
| MW | A measure of power. It is a million watts. Submultiples can be kW (kilo-watts) or W (watts). |
| MWn | Nominal MW. This refers to the nominal power in watts of a given pho-tovoltaic installation. The maximum power that a solar PV feed-in type installation provides at the output of an inverter. |
| MWp | The peak power as measured in MW as a consequence of the addition of the peak power of each of the solar PV modules of a given installation, as specified by the manufacturer (usually, when the module surface is |

perpendicular to the sun, with 850–1,000 watts/m$^2$ and at 20–25 °C of ambient temperature, depending on the module specs).

MPPT    Maximum power point tracker. Is a technique that grid-tie inverters, solar battery chargers and similar devices use to get the maximum possible power from one or more photovoltaic devices, typically solar panels.

O&M     Operation and Maintenance of a given system.

OMEL    Operador del Mercado Eléctrico (http://www.omel.es/inicio). It is the entity managing and supervising the correct economic and administrative development of the electricity market in Spain.

OMIE    Operador del Mercado Ibérico de Energía polo Español. It is the new OMEL that supersedes to the OMEL but for the Iberian market.

PCS     Power conditioning system.

PR      Performance ratio. It refers to a product's ability to deliver performance. In the case of solar PV systems, it specifically refers to the relation between actual yield, considered as the energy delivered by the sun on the PV modules, and target yield of a PV system, after having discounted usual losses. A photovoltaic system with a high efficiency can achieve a performance ratio over 70 %.

PV      Photovoltaic.

R&D     Research and development.

REE     Red Eléctrica Española (www.reee.es). It is the electric system operator, which guarantees the continuity and security of the power supply and the proper coordination of the production and transmission system in Spain.

sc-Si   Single crystalline, also known as monocrystalline silicon, a technique of PV cells for solar modules.

SPE     Special purpose entity. A legal entity (usually a limited company of some type or, sometimes, a limited partnership) created to fulfill narrow, specific, or temporary objectives.

SPV     Special purpose vehicle. See SPE above.

Toe     Ton of oil equivalent, a measure of energy equal to 42 GigaJoules.

TEU     Twenty-foot Equivalent Unit and is a standard measure for containers' lenght.

UNESA   Asociación Española de la Industria Eléctrica. It coordinates, represents, manages, promotes, and defends the interests of the associated electrical companies.

UMTS    Universal mobile telecommunication system. A third-generation mobile system.

WEO     World energy outlook. An annual report released by the International Energy Agency.

# Chapter 1
# Introduction: Solar Energy and Human Civilization

Solar energy has made life possible for humans and all other living things on planet Earth. Since the industrial revolution began in the nineteenth century, humanity has increasingly become dependent on a second source, solar energy-derived fossil fuels (coal, oil, natural gas). The quantity of energy from the combustion of fossil fuels surpassed that derived from biomass during the 1890s (Smil 1994). At that time humans consumed about 600 billion watts (GW) of fossil fuels per year, and by 2005 the world was consuming fossil fuels at the rate of 12 trillion watts (TW), a 20-fold increase (Smil 2006). This unprecedented increase in the availability of energy-dense fossil fuels has enabled the exponential growth of the human population and affluence for an unprecedented number of people (Hall and Klitgaard 2011).

Fossil fuels in general are finite resources subject to depletion (Hubbert 1969; Campbell 1998). Currently, more than 50 oil-producing countries have experienced a peak (i.e., the point of maximum production) in daily oil production. Peak oil for the entire world is thought by many to be happening about now, and peaks in other fossil fuels will follow in the not too distant future (Fig. 1.1). Since anyone reading this book has lived almost all of his or her life during an era of cheap, abundant fossil fuels, the idea of energy-supply constraints on economic growth may be hard to accept. But the concepts and the data to support this relationship are fairly straightforward, especially for oil, which is our most important fossil fuel. Today, we are consuming oil at two to four times the rate at which we are finding it. We know this is true because for the past four decades there has been a substantial decline in the rate of discovery of new oil fields in spite of ever increasing drilling efforts. Moreover, the decline rates of the largest oil fields, which supply 80% of our oil today, are alarming. They range from 2% to almost 20% yearly, with an average decline rate of 6.7%. This average decline rate is expected to increase to 9% per year if future investments in oil exploration are not large enough and perhaps even if they are (Simmons 2007; Birol 2008).

P.A. Prieto and C.A.S. Hall, *Spain's Photovoltaic Revolution: The Energy Return on Investment*, SpringerBriefs in Energy, DOI 10.1007/978-1-4419-9437-0_1, © Pedro A. Prieto and Charles A.S. Hall 2013

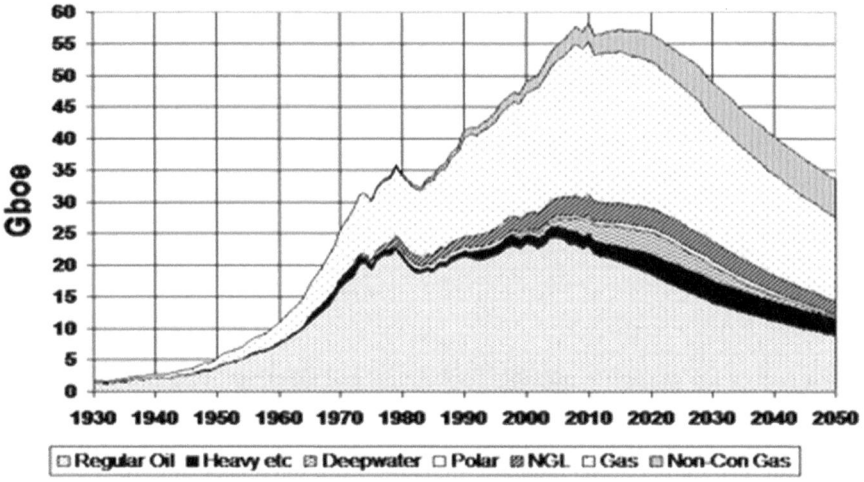

**Fig. 1.1** History and projection of main fossil fuels for global human society. Peak "regular" or conventional oil is thought to have occurred around 2005 although various nonconventional "oils" (such as bitumen from "heavy" tar sand deposits) have delayed the peak of total liquids. Natural gas is still expanding but will have its own peak in the next decade or so (courtesy of Colin Campbell)

Peak oil presents society with some extremely difficult challenges. As energy analyst Vaclav Smil has put it:

> Speaking as a scientist relying on first principles, I must stress the extraordinary scale of the coming transition: the shift to non-fossil energies is an order of magnitude larger task than was the transition from phytomass to fossil fuels, and its qualitative peculiarities will also make it more, rather than less, demanding; consequently, its pace may have to be much slower than is commonly assumed. At the same time, only one of the many renewable energy resources has a natural flux far surpassing any prospective needs, and the remaining palette of available choices is not at all as bright as seen through the eyes of true believers (Smil 2006).

Smil believes that direct sunlight is the only source of energy capable of providing the necessary energy for today's human civilization. Once fossil fuels have peaked globally and begin their steep decline, humanity will be faced with a diminishing supply of cheap and high-density energy in the range of 10–20 EJ per year, which will be a significant part of the 400 EJ we are using today worldwide. For the past 6 years, global oil production and total net energy use has not been growing, which certainly has contributed to many of the economic problems facing the world today. This is such an important issue that society should be seriously focusing on moving toward alternative energy sources to replace oil, something that has not yet happened at a global scale.

# Why EROI?

But there remains at least one very large barrier to moving smoothly to solar energy: it is monetarily very expensive, and where something is expensive in dollars, it will be expensive in energy. This applies especially to the investments required to generate any substantial amount of energy. There are several approaches to understanding the relation of energy invested and energy gained, and the one we very much prefer is called energy return on investment, or EROI, or sometimes energy returned on energy invested, or EROEI. This concept applies as well to individual survival, so we will begin from a biological perspective.

A universal property of all life is the need to invest some of the energy obtained into all of life's functions, including that of obtaining more energy. For example, if the cheetah is to survive as a species and propel its genes into the future, the energy gained by the average cheetah eating a gazelle in the African savannah must be substantially greater than the energy used to run down the gazelle. The cheetah must also be able to fuel its own maintenance metabolism (the energy its body requires to maintain essential physiological functions), plus the additional energy for hunting attempts that were unsuccessful and reproduction and nurturing of offspring. In other words, the cheetah must have a substantial energy return on the energy invested (EROI) in hunting gazelles or other prey to account for all of its natural requirements for life (Hall and Klitgaard 2011).

Our preagricultural hunter–gatherer ancestors had, such as we can understand from their modern counterparts, an EROI on the order of 10:1, which is high enough to survive, reproduce, and have the necessary leisure time for a healthy life, but not enough to generate a very large surplus (Hall and Klitgaard 2011). This explains why hunter–gatherers created relatively few energy-expensive monuments, such as the post-agricultural Egyptian pyramids, Stonehenge, or the Great Wall of China, nor did they change their environment in any significant way. The discovery of fire, the domestication of animals, and especially the development of agriculture incrementally added the necessary net energy needed for human civilization, as we know it, to evolve and expand, and began the transformation of the Earth's surface through agriculture, urbanization, etc.

Domesticating animals for milk and meat and agriculture gave humans more net energy than hunting and gathering for a whole range of activities. This energy surplus gave way to the division of labor, which included ironsmiths, craftsmen, traders, and so on. This in turn increased the ability of human societies to generate energy surpluses to support cities, artisans, the military, and the ruling classes, which included monarchies, priests, and specialized accountants. When human societies were able to generate great surpluses of energy over their strict metabolic needs, they often utilized that surplus energy to expand their civilizations. They accomplished this by building magnificent structures such as the Parthenon, ziggurats, pyramids, irrigation channels, palaces, and walled cities. These structures gave them the symbolic power that they needed to absorb surrounding, weaker societies. Of course, they often used their surplus energies in military adventurism, which sometimes helped them to obtain the surplus energy of other societies (such as land, the main solar collector of the past) and sometimes simply wasted their surplus energy in vain. Many have told the stories of how human societies have risen and fallen in connection to energy elegantly and in detail, but the best one to our knowledge can be found in Joseph Tainter's *Collapse of Complex Societies* (Tainter 1988).

As time progressed, developments in technology have allowed humans to exploit increasingly large energy resource bases. This technological evolution began with hunter–gatherers' use of their own mechanical energies and evolved into complex human societies' utilization of slaves, serfs, and draft animals to exploit the concentrated solar power stored in cultivars (agriculture) and then to modern human societies' utilization of biomass, wind, water, and solar power and to today's use of fossil fuels. Until about the 1800s essentially all the energy used by human societies was solar-based and taken from the biosphere, basically a two-dimensioned environment. The jump to the use of energy from the lithosphere (first fossil fuels and later uranium) required energy-intensive technologies for the exploitation of this third dimension. This process increased the energy investment costs, but the return was proportionally more substantial. In the meantime, the increasing power in the hands of humans also initiated the depletion and eventual exhaustion of some renewable resources such as fish and biomass in virgin forests.

## Solar Energy-Derived Fossil Fuels

Coal, the first fossil fuel to be developed and exploited, gave humans an enormous advantage in the energy available, in the intensity of use, and therefore in the potential energy flows that could be supplied to society. Coal required biomass to be exploited in the form of underpin beams, railway sleepers, housing beams, barrels, wood containers, etc., and a considerable use of human and animal mechanical energy.

After the American discovery of the Spindletop oil field in Texas in 1901, oil and subsequently the total energy production of the world were vastly increased, especially after the development of rotary drilling. This represented a new kind of energy investment that for most of the twentieth century had extremely large returns (Hall and Klitgaard 2011). Throughout the 1900s, oil took over the role of coal and biomass in many applications because of its unique qualities, which included its versatility, high energy density, and transportability. Internal combustion engines (ICEs), powered by oil derivatives, e.g., kerosene and later gasoline soon were helping to mine, extract, and process coal or, using power saws, cut wood from forests. Coal, too, helped in the manufacturing of steel pipes for exploration, drilling, or transportation of oil and its derivatives through pipelines or oil tankers. This energy interdependency grew but obviously in an asymmetric form, with the most powerful, dense, versatile, and convenient fuel (oil) taking the lead and replacing many of the functions of the previous ones (coal and wood) (Fig. 1.2).

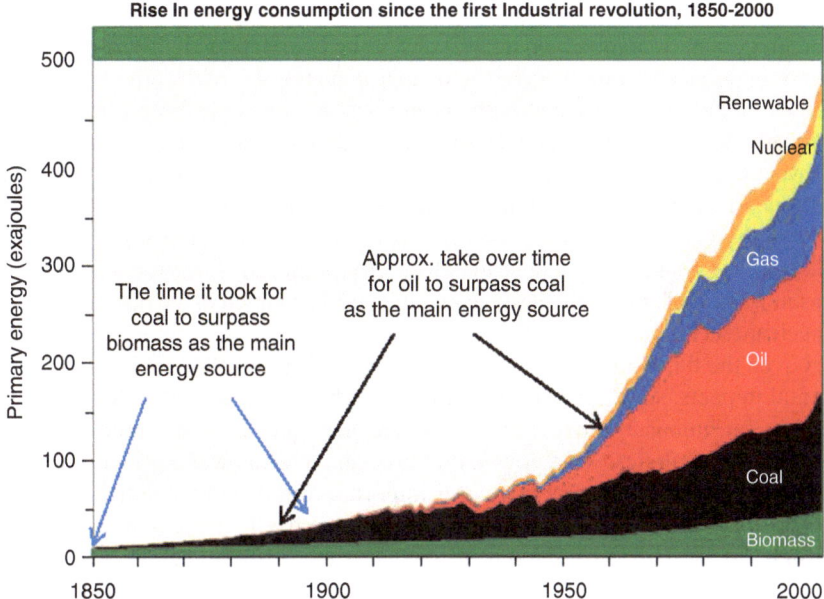

**Fig. 1.2** World's energy consumption since the industrial revolution has paralleled the increased human economic development and population growth for the past 161 years. Renewable energy is mostly hydroelectric power. Modified from World Economic and Social Survey, UN, 2011

While oil was never easy to find or extract, the quantity available increased reliably at 3% a year for the first half of the twentieth century. Since 1950 the rate of increase began to decrease and now oil use appears to be no longer increasing at all. This phenomenon, known as peak oil, is a geological fact for the overwhelming majority of oil-producing countries even with many technical improvements in oil drilling. An additional problem is that the EROI for oil tended to be higher in the past compared to recent decades. From 1901 through the 1930s, oil discoveries frequently gave an energy return of 100–1,000 barrels per barrel spent seeking it. The extraction of the first, sweet, easily accessible, and abundant oil in the United States had a very high EROI of 20–30 or more barrels returned per barrel of oil, or its energy equivalent as gas invested in producing it (Guilford et al. 2011). These values have been decreasing to less than ten returned to one invested today, as the largest and most accessible deposits are being exhausted and as the industry must turn to deeper, heavier oil in smaller, less accessible oil fields. This reality has forced oil companies to spend more energy to obtain and refine the same amount of oil, thus lowering its EROI. Similar patterns of decline are occurring around the world although at a rate usually delayed relative to the USA where the modern oil industry was born and depletion of conventional sources is most complete.

Other energy sources, such as hydroelectric and nuclear power, have contributed to shape the world as we know it today. But they never surpassed fossil fuels in quality and quantity, as coal did once with biomass and later oil with coal. Therefore, it is important to consider both the quality and the quantity (actual and maximum potential) when we embark on an analysis of the potential of any source of energy. Today, hydroelectric power accounts for approximately 4% of the primary energy of the world, and its further development in most countries has been limited due to relatively few remaining suitable sites and major environmental considerations. Conventional nuclear-generated electricity has been operational for more than half a century, fueled by finite quantities of uranium. Nuclear power represents less than 6% of the world's total primary energy (Fig. 1.2). This number drops by two thirds when we only take into account the electricity produced, since for some peculiar reason waste and environmentally damaging heat is counted in the official output of a nuclear reactor.

Given the finite nature of fossil fuels, their declining EROI, and their environmental impacts, such as their production of carbon dioxide ($CO_2$), there are good reasons for human civilization to turn to cleaner solar-derived sources of energy such as photovoltaics, wind power, biomass, and hydropower even when they are not so economically competitive. An important question is whether renewable sources of energy, when they are not being subsidized, can deliver enough surplus energy to society for them to be competitive with "depletable" and perhaps dangerous fossil fuels. But what if the EROIs of these alternative energy sources are small compared to the fuels they are replacing? And what would too low mean to society?

## What Level of EROI Do We Need?

Unlike our cheetah, we need more than a "just barely sufficient" EROI to survive if the net profit is to support our complex society. Quoting from Hall (2011): "Think of a society dependent upon one resource: its domestic oil. If the EROI for this oil was 1.1:1 then one could pump the oil out of the ground and look at it. If it were 1.2:1 you could also refine it and look at it, 1.3:1 also distribute it to where you want to use it but all you could do is look at it. Hall et al. 2008 examined the EROI required to actually run a truck and found that if the energy included was enough to build and maintain the truck and the roads and bridges required to use it (i.e., depreciation), one would need at least a 3:1 EROI at the wellhead. Now if you wanted to put something in the truck, say some grain, and deliver it, that would require an EROI of, say, 5:1 to grow the grain. If you wanted to include depreciation on the oil field worker, the refinery worker, the truck driver and the farmer you would need an EROI of say 7 or 8:1 to support their families. If the children were to be educated you would need perhaps 9 or 10:1, have health care 12:1, have arts in their life maybe 14:1, and so on. Obviously to have a modern civilization one needs not simply surplus energy but lots of it, and that requires either a high EROI or a massive source of moderate EROI fuels" (Fig. 1.3).

Humans currently consume approximately 12.5 trillion watts (TW) of industrial power (i.e., energy delivered by human industry as opposed to the sun that runs

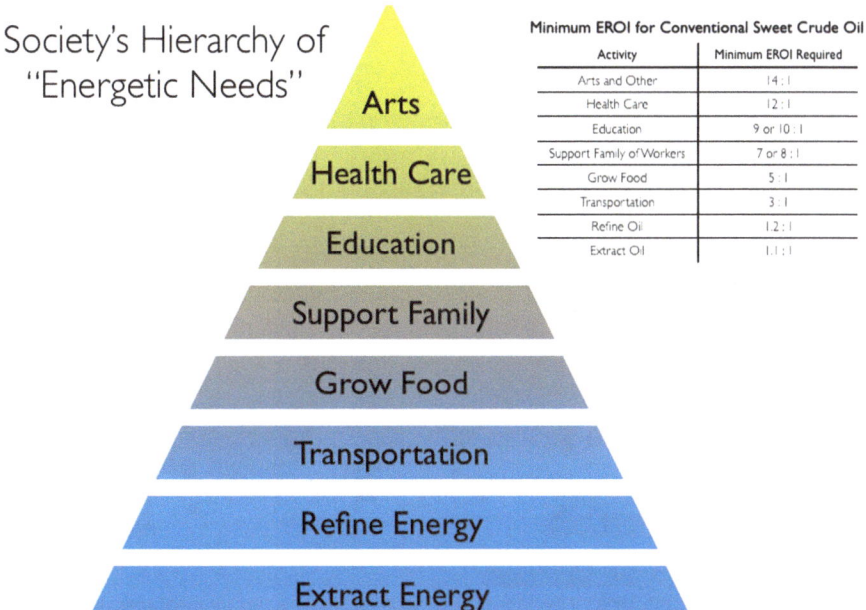

Fig. 1.3 "Pyramid of energetic needs" representing the minimum EROI required for conventional oil, at the wellhead, to be able to perform various energetic task required for civilization. The *blue* values are published values: the *yellow* values are increasingly speculative (figure adapted from Lambert and Lambert 2012)

natural ecosystems and agriculture), and this power is being generated primarily (approximately 86%) from fossil fuels (Jacobson and Delucchi 2011). Can we possibly replace fossil fuels with renewable sources of energy? Solar photovoltaic energy should have a high potential for replacing fossil fuels as the EROI of the latter continues to decline in the future. The amount of solar energy hitting the Earth each year is approximately 6,500 TW. One estimate is that humans could harness approximately 360 TW of the solar energy that strikes the continents of the Earth with photovoltaics alone (Jacobson and Delucchi 2011). But there is a catch—solar radiation is diffuse and intermittent and is not easy to capture without a great investment from the main industrial society, which of course is run mostly on fossil fuels.

## Why the Kingdom of Spain?

Spain, located on the Iberian Peninsula in southwestern Europe (Fig. 1.4), is the sunniest country in Europe and has been at the forefront of investing in solar photovoltaic (PV) technologies as a major component of its energy policy.

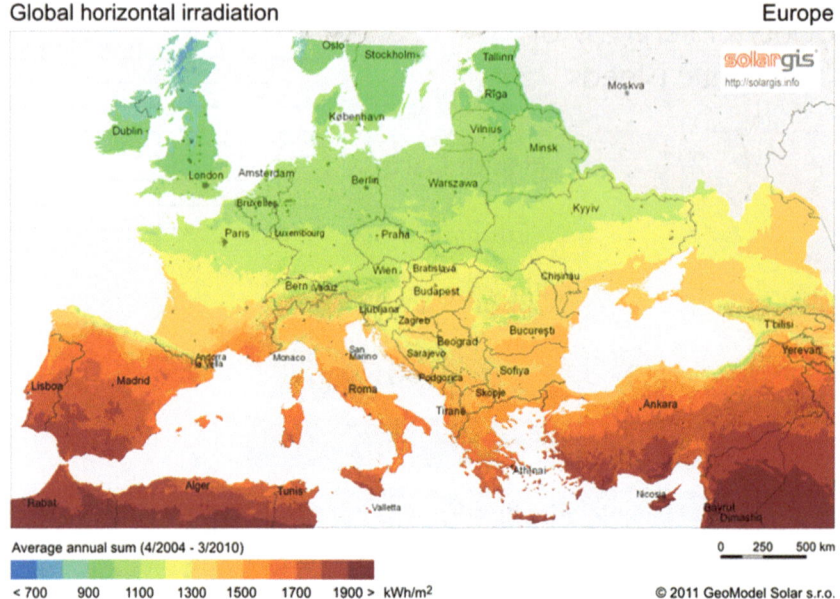

**Fig. 1.4** Horizontal irradiation in Europe. *Source*: Solargis. http://www.solargis.info

US President Obama has referred to Spain several times as an example to follow when it comes to its renewable energy policy. In the past decade, Spain has become the second largest country in the world for installed photovoltaic power with a total installed capacity at the end of 2011 of 4,237 megawatts (one million watts, MW) (as a point of reference, 1,000 MW is the size of a large coal or nuclear plant). Spain's privileged geographical location on the sunny Iberian Peninsula, its relatively sophisticated industry, and its many years of massive photovoltaic deployment in full operation make Spain the ideal country in the world for examining EROI of photovoltaics. While Germany has more installed capacity (Fig. 1.5) and currently leads the world in the investment of photovoltaic technologies, Spain's solar energy generation potential and EROI should be much higher than that of the country with the largest PV installations, cloudy Germany.

Our analysis begins with a qualitative and quantitative overview of the Spanish solar experience in recent years. This includes the growth of primary energy consumption to feed the country's electrical network, the evolution of photovoltaics in Spain and the bureaucracy regulating it, the structure and functioning of the Spanish electric grid, and an overview of the impressive growth of the renewable energy sector in Spain, which has prompted many other countries around the world to look at Spain as an example to follow.

We first take a look at the policies and tariffs that Spain has implemented to encourage solar PV, the country's learning curves with these policies, and the interaction of all of these issues with the domestic and global solar PV markets in terms

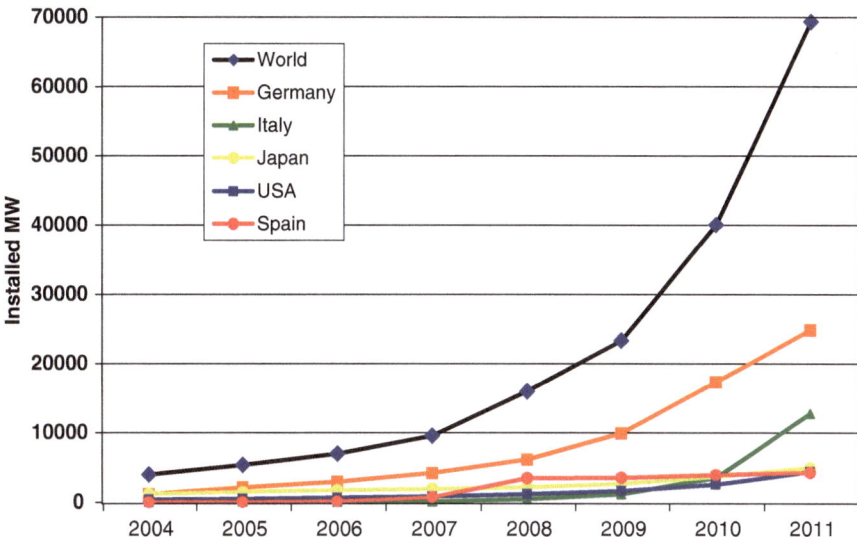

**Fig. 1.5** Installation of solar power in the five countries with the largest installed capacity (in potential MW at full sunlight). These countries together generate 80% of the total PV energy for the globe. It should be noted that in each country these values are, at best, in the range of 1% of all energy use and usually not more than 3% of total electricity use. *Source*: BP Statistical Yearbook 2011 for data up to 2010. Also from IEA Photovoltaic Power Systems Programme, EPIA, EurObserver and SolarBuzz

of prices as well as the quantity and the quality of the electricity delivered to Spaniards. We consider the plans for the future of renewable energies in Spain and the dilemma the Spanish government is facing today with regard to their commitment to renewable energies. In particular, we examine the contribution of the deficit derived from their need to honor the premium feed-in-tariffs (FITs) (i.e., government subsidies) amidst the Spanish and global financial crisis and the necessary financial tightening and forced readjustments of the Spanish government. We discuss whether Spain can reach grid parity (i.e., solar energy costing the same as electricity in general) and also how the original rules were bent by financial greed and even, possibly, fraud in some installations during the solar PV revolution of Spain.

## Spain (2009–2011): An Excellent Test Bed for Solar PV Energy Costs and Gains

Spain is an excellent country to undertake an EROI analysis of solar PV because it has an efficient national accountability system for solar PV electric generation and operation. Madrid has created such a system in order to be able to register and supervise the country's solar PV-installed power and power generation. The government needed an up-to-date and accurate measure of all the energy generated with renewables in order for it to administer the premium tariffs it distributes among generators. All power plants considered here have been registered with the government, and their generation is constantly reported, on a monthly basis, to the governmental institutions that keep track of these data and make it public on official governmental web pages.

The production of electricity from solar PV was quite stable and well measured during 2009, due to the minimal growth during that time period (Fig. 1.4). This has helped us to analyze Spain's solar PV with considerable precision. We also include 2010 and 2011 in some analyses. Although there was more growth in electricity generation than in 2009, output was still relatively stable. Finally, Spain is behaving very much like an island with respect to electrical generation, with little transfer across its borders. These characteristics helped us to analyze the possible upper limits of the intermittent energy that can be injected into the national grid from photovoltaic devices.

## Energy in Spain

Nature did not endow Spain with many fossil fuels, except for several deposits of low-quality brown lignite coal, which presently contribute one third of the total national coal consumption. The other two thirds are imported, mainly from South Africa and Australia.

Originally, Spain was fully covered by energy-rich hardwood forests, as mentioned in Roman literature. These forests were preserved up to at least the twelfth century. The Spanish discovery of America in 1492 gave way to the age of colonialism. This new age required long-haul sailing ships (Caravels) to travel across the Atlantic. Consequently, Spain built a powerful fleet in order to conquer and eventually trade with the American continent. This provoked a deforestation that completely changed the landscape of the Iberian Peninsula but helped Spain to become the first world power. In the early stages of Spanish colonialism, Emperor Philip II began to have some concerns about the massive use of wood for construction and shipbuilding. Phillip II and his successors wasted most of the Spanish biomass energy on wars, the most famous being the Spanish Armada and its failed attempt to conquer England in 1588. After Napoleon conquered Spain (from 1808 to 1814), the Spanish Empire lost most of its colonies in the New World and began to decline as a world power (Fig. 1.6), although to some extent its forests recovered. Today, 31% of Spain is covered by forests but it remains a country with a mostly dry climate, except in the northern strip of the Peninsula, where it rains more than in many other countries of Europe.

The industrialization of other parts of Europe took place earlier than in Spain. The civil war (1936–1939) added another important delay in the process of industrialization. The Francisco Franco dictatorship (1939–1975) initiated the construction of a number of dams throughout the country, both for irrigation and electrical generation, although their potential output was limited due to Spanish orography and the limited rainfall on most river basins. Today, hydroelectric generation, mostly from the hydropower plants of that period, meets about 10% of the total electricity demand in Spain.

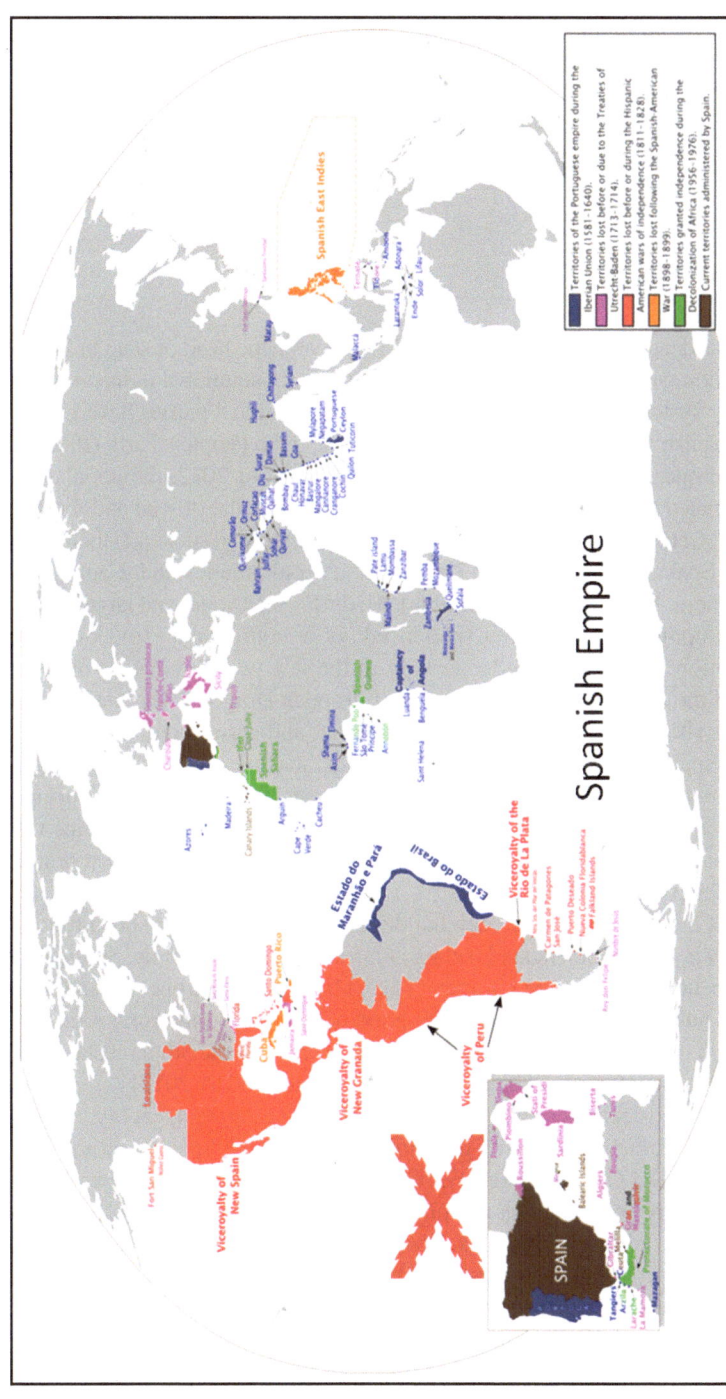

**Fig. 1.6** The Spanish Empire. This empire required massive amounts of energy resources, mainly in the form of biomass, in order to create and maintain territories across the world. *Source:* Wikipedia contributors. "Spanish Empire." Wikipedia, The Free Encyclopedia. Wikipedia, The Free Encyclopedia, 20 May. 2012. Web. 20 May. 2012

Spain began its serious industrialization in the late 1960s with the advent of the "technocratic government" during Franco's regime. This led to the development and construction of eight nuclear power plants, although a couple of them were never commissioned due to security and economic reasons. Before and especially after joining the European Union in 1986, Spain experienced a continuous growth in its appetite for energy as its population and economy expanded.

## Contemporary Spain

Since Franco's death in 1975, King Juan Carlos I became the head of state in Spain. A new constitution was approved in 1978, establishing a constitutional monarchy with a parliamentary system. In the past 30 years, two main political parties have governed Spain: the Spanish Socialist Workers Party (PSOE) and the People's Party (PP). Spain joined the European Economic Community in 1986. As of 2012, the population of Spain is estimated to be 47,000,000 (CIA 2012). Spanish is the official and most spoken language. Catalan, Galician, Valencian, and Basque are also spoken in their respective regions. Spain is composed of 17 autonomous communities and 2 autonomous cities in Northern Africa (Ceuta and Melilla). Madrid is the capital and largest city.

The development that Spain experienced as a consequence of joining the European Union was impressive, especially in the last two decades during which Spain earned the right to be called a sort of "European Dragon." The main Spanish industrial activities are in the tourism, automaking, and agricultural sectors. In the last decade, Spain experienced a real estate boom, starting more housing developments than most of the other countries in the European Union combined. During the past three decades, Spain built one of the largest automaking industries in the world with its famous brand SEAT. As of 2012, SEAT is the world's ninth largest producer of automobiles (OICA 2012).

In the past 5 years, Spain has been building an impressive high-speed train system called AVE, with sophisticated technical infrastructure. Currently, Spain has the highest rail penetration rates in kilometers per inhabitant in Europe. Spain has also invested in a national electricity grid with good penetration to the rural regions. As of 2012, Spain is the 12th largest economy in the world and the 5th largest economy in the European Union with a GDP of 1.05 trillion Euros (CIA 2012).

In the past three decades of very rapid economic development, energy consumption grew at about the same rate as the economy (Fig. 2.1). Spain had, however, a much higher level of $CO_2$ emissions per capita than many of its European counterparts. With this large economic development, Spain exceeded the established target in the Kyoto Protocol by nearly 150%. The types of industries in Spain have needed to consume more fossil fuel energy per unit of economic product produced than those in more developed countries which have tended to outsource the most energy-intensive (and hence polluting) industrial activities.

The world financial crisis of 2008 has impacted Spain severely, especially in the real estate sector, where the bubble created in the past decade burst. As a consequence of the previously held firm belief in perpetual economic growth, nurtured by many well-meaning economists and politicians, this economic system collapsed along with Spain's economy. Today, Spain is suffering from an unprecedented economic crisis mainly caused by the bursting of the housing bubble in 2008. Curiously, however, its public debt (until the last few years) was relatively small due to an efficient government that managed its economy relatively well.

Solar photovoltaic energy contributed to these problems because it has costs as well as gains. This is why we believe it is important to analyze both the benefits and the costs of this technology in order to derive the relation of the outputs from the Spanish solar program to its substantial input cost through the use of a comprehensive EROI analysis.

The purpose of this book is to analyze whether Spain's photovoltaic revolution has had an energy return on investment capable of capturing and transforming enough solar energy for modern human civilization. Can renewable energies replace, in a relevant timeframe, fossil fuels, at least partially and at something like the energy levels the world consumes today and in time to stabilize climate change? We will examine that question at the national level by attempting to account for all of the investments required to supply photovoltaic energy to an economy.

# References

Birol, F. 2008. *World energy outlook*. Paris: International Energy Agency. http://www.iea.org/text-base/nppdf/free/2008/weo2008.pdf. Accessed 19 Sep 2012.

Canadian Solar. 2012. http://www.canadiansolar.com/en/products/product-documentation/product-documentation.html. Accessed 19 Sep 2012. See warranty zip or pdf.

Central Intelligence Agency (CIA). 2012. *The world factbook: Spain*. Updated 10 September 2012. https://www.cia.gov/library/publications/the-world-factbook/geos/sp.html. Accessed 15 Sep 2012.

Gagnon, N., C.A.S. Hall, and L. Brinker. 2009. A preliminary investigation of energy return on energy investment for global oil and gas production. *Energies* 2(3): 490–503.

Hall, C.A.S. 1972. Migration and metabolism in a temperate stream ecosystem. *Ecology* 53(4): 585–604.

Hall, Charles A.S. 2011. Introduction to special issue on new studies in EROI (Energy Return on Investment). *Sustainability* 3(10): 1773–1777.

Hispanidad. 2010. *Las subvenciones a las renovables destrozan el ciclo combinado*. http://www.hispanidad.com/noticia.aspx?ID=133710. Accessed 19 Sep 2012.

Instituto para la Diversidad y Ahorro de Energia IDAE 2011. *Plan De Energia Renovables 2011–2020*, 147 (Table 10.a). http://www.idae.es/index.php/mod.documentos/mem.descarga?file=/documentos_11227_PER_2011-2020_def_93c624ab.pdf. Accessed 19 Sep 2012.

Organisation Internationale des Constructeursd' Automobile (OICA). 2012. *World motor vehicle production ranking by country 2010–2011*. http://oica.net/wp-content/uploads/total-2011-august-2012.pdf. Accessed 15 Sep 2012.

Simmons, Matthew. 2007. PEAK OIL: is it real? When might it occur? In *Kayne Anderson energy funds partners' meeting*, Houston, 12 February 2008 (Chart 35).

Smil, V. 1994. *Energy in world history*. Boulder: Westview Press.

Smil, Vaclav. 2006. Energy at the crossroads. Background notes for a presentation to the Global Science Forum Conference on Scientific Challenges for Energy Research, Paris, May 17 and 18.

Tainter, J. 1988. *The collapse of complex societies*. Cambridge: Cambridge University Press.

# Chapter 2
# The Evolution of the Demand for Primary Energy and Electricity in Spain

In the past, electricity production in Spain was most dependent on coal, but it transitioned quite rapidly to gas after building the Pedro Duran Farell gas pipeline (1996) connection with Algeria through Morocco and then a second pipeline directly from Algeria (Fig. 2.1). Spain has the largest liquefaction and regasification network port terminals in Europe to produce and transfer liquid natural gas (LNG). Natural gas was implemented rapidly for several reasons: (1) the Spanish government attempted to comply with the Kyoto Protocol by trying to decarbonize, (2) it was accessible in huge amounts via pipeline from Algeria, and (3) Spain has the best regasification infrastructure in ports and one of the best LNG tanker fleets in Europe to bring it from gas producers in Africa. The result has been an increasing dependence on imported oil, gas and a continued, although lessened, use of imported coal, along with a resultant drain of foreign exchange used to pay for the imported fossil fuels (Fig. 2.2).

## Recent Evolution of the Spanish Electricity Supply

Spain first implemented a special program to develop renewable energies in 2004. Given that most of Spain is a dry, even semi-desert country, with extremely abundant sunshine, it would seem to be one of the most reasonable places in the world to develop an extensive solar power system (Fig. 2.3). This would require a substantial program integrating engineering, grids, financial incentives and good policy. Consequently, the Spanish government started a very large program with the objective to reduce Spanish dependency on fossil fuels and to minimize the foreseeable economic impacts of meeting Al Gore's Kyoto Protocol on climate change. Spain basically followed the German "feed-in-tariff" (FIT) policy approach to renewable energies. This policy is based on the government requiring utilities to provide renewable generators with a long-term fixed price for electricity, based on either a system's generation costs or fixed premium tariffs paid on top of the spot market

P.A. Prieto and C.A.S. Hall, *Spain's Photovoltaic Revolution: The Energy Return on Investment*, SpringerBriefs in Energy, DOI 10.1007/978-1-4419-9437-0_2, © Pedro A. Prieto and Charles A.S. Hall 2013

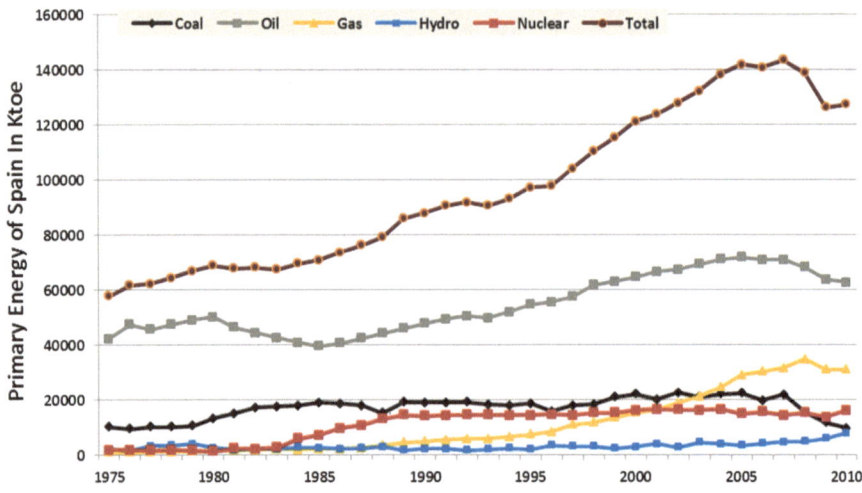

**Fig. 2.1**  Evolution of primary energy generation in Spain (1975–2010). *Source*: La energía en España (2010). Ministerio de Industria, Turismo y Comercio, p. 333). This source does not include renewable energies outside of hydropower (1 kilotonne oil equivalent = 42 terajoules)

## 2010 Primary Energy Consumption in Spain: 132.1 MToes

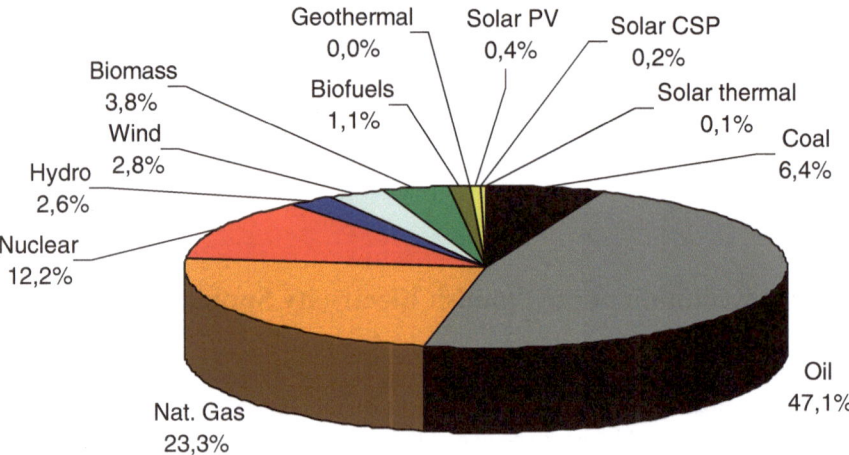

**Fig. 2.2**  Demand/consumption of primary energy in Spain in 2010. *Source*: Ministry of Industry, Tourism and Commerce, p. 37 (131.2 MToe = 5.5 exajoules) and percentages of each energy source

price for electricity. This policy has encouraged investment in renewable energy technologies such as photovoltaics through long-term subsidies (Fig. 2.4).

Spain had a complex mix of electricity production (Fig. 2.5), which is essentially the same as demand or consumption. During this year, 30% of the electricity was generated from renewable sources of energy, or as much as 44% of the electricity if

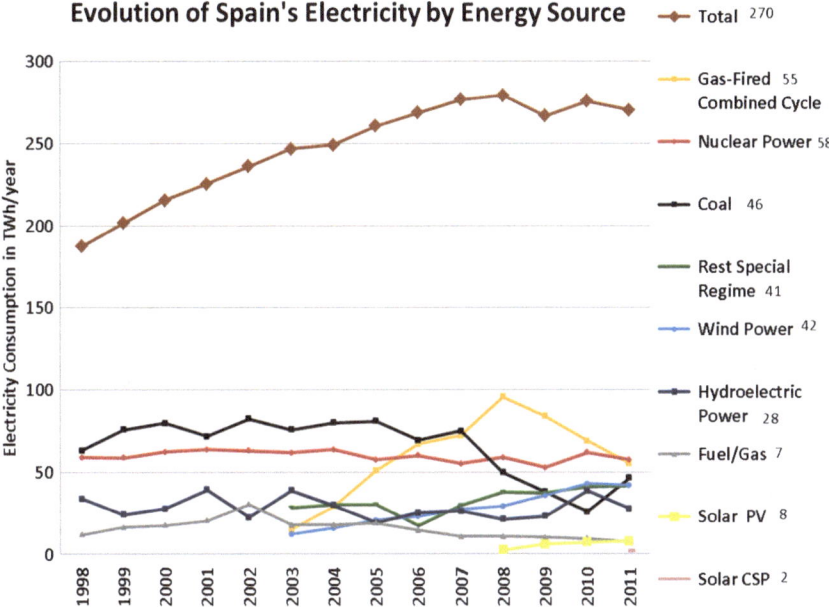

**Fig. 2.3** Evolution of the electricity supply in Spain by energy source (1998–2011). These data do not include international exchanges or energy costs for pumped storage or consumption by the supply industries. The "special regime" includes biomass in different forms and electricity generated from waste and cogeneration. *Source*: Red Electrica Española (REE)

cogeneration and waste treatment are included. The feed-in or entry criteria of the different sources are rather complex, but in general terms, renewables have priority of entrance into the grid.

## An Overview of the Status of Renewable Energies in Spain

In 2004, the Spanish government introduced the "special regime" to encourage with subsidies (called premium tariffs) low-$CO_2$ generators, e.g., biomass, solar PV, CSP, wind and cogeneration. Investors started to develop renewable energy technologies in Spain in "feed-in" form (meaning that each would be feeding electricity into the national grid). Soon, Spain became a mature country in this type of connected and distributed electricity generation and network system. Unfortunately, that soon created problems of network capacity in some regions where the power lines or electric substations did not have enough capacity to absorb this type of stochastic (intermittent) energy. Spain had an impressive learning curve under the hand of the Spanish regulators and the main electric power utilities. They managed to overcome most of these problems by combining their expertise.

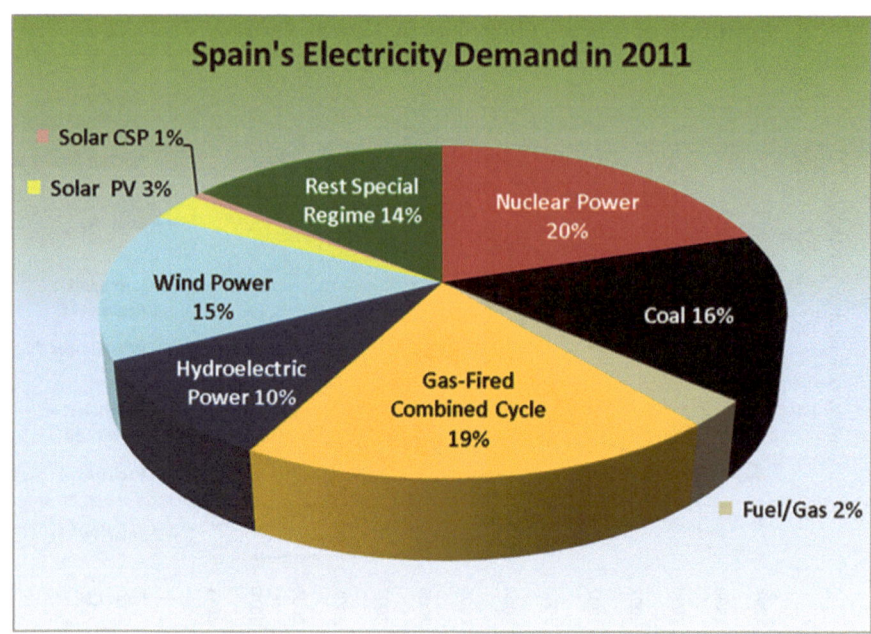

**Fig. 2.4** Electric consumption in Spain in 2011. *Source*: Red Eléctrica Española. Preliminary Report 2011

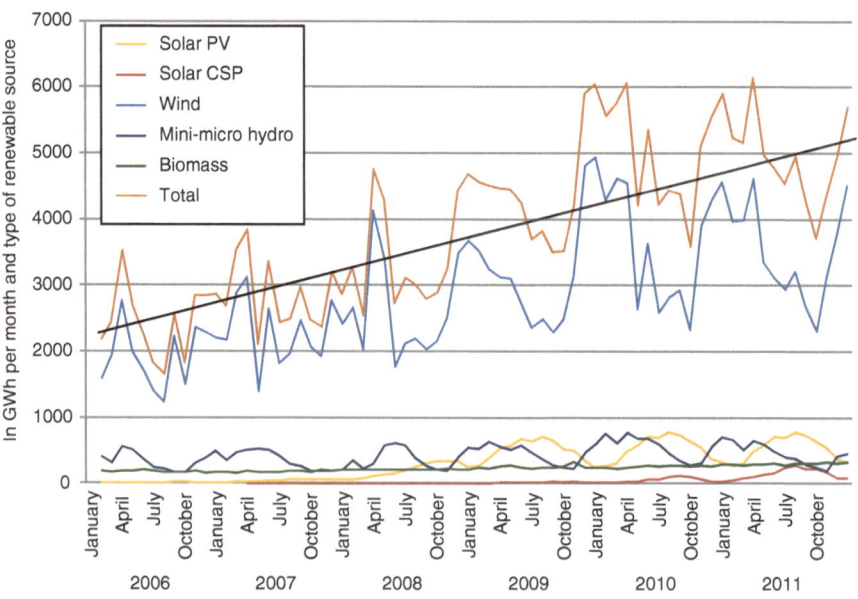

**Fig. 2.5** Evolution of feed-in renewable energy expressed in GWh per month in the last 6 years. *Source*: Comision Nacional de Energía. Report of April 2012. Installed power at the end of 2011 was 29,047 MW of these types of renewable sources (27% of total electricity produced that year; www.ree.es. Preliminary report 2011

   Three unusual characteristics of Spain with respect to the introduction of renewable, distributed energies are (1) its feed-in system; (2) its unique geography, in the southernmost part of Europe and on a Peninsula; and (3) very few connections and exchanges of electricity with neighboring countries. These characteristics allow Spain considerable autonomy and flexibility in developing its own renewable energy systems. The few international exchanges with France, Portugal, and Morocco are small. Although Spain has been a net exporter of a relatively small amount of electricity in the last few years, it has done so with a very low exchange capacity, due to lack of international interconnections. All of these factors make Spain a unique and interesting location to undertake a thorough EROI analysis of any renewable energy.

# Chapter 3
# The Historical, Legal, Political, Social and Economic Context of Solar Photovoltaics in Spain

The main factors behind the explosive growth of PV installations in Spain are as a consequence not of biophysical but of social factors, which we review here.

## Development and Deployment of Photovoltaic Energy Technologies in Spain

The progress of Spain's solar PV technologies and their deployments has been impressive in recent years. Spain followed the example of Germany's policies to incentivize the private investment in solar PV technologies, which overall have been a success. In 2009 and 2010, Spain was the second largest producer of photovoltaic electricity in the world after Germany. By the end of 2010, Spain had reached 10% of the world's PV-installed power (Fig. 1.5). In 2009, solar PV provided 2.26% of the national electricity demand, the largest penetration of solar PV of any national electrical grid in the world, with peaks as high as 5% in June of 2011.

Estimating solar PV electricity production is usually difficult where installations are added throughout the year while connected to an international grid. However, the collapse of fast-paced PV installations in Spain in 2009, declining from 2,708 MW of new installed capacity in 2008 (the world record, with more than half of the world's installations that year) to not one single official MW in 2009 (Table 3.1), and very few MW in 2010 and 2011 makes Spain in 2009 through 2011 a unique reference point to assess with great accuracy how much electricity was generated from solar PV plants. PV-installed power was relatively constant throughout 2009, with only a small growth in 2010 and 2011 that allows for an easy extrapolation of the average installed base for Spain (Table 3.1). There is no other country in the world with such accurate solar photovoltaic statistics from various official sources accurately and professionally tracking a variety of data sets.

P.A. Prieto and C.A.S. Hall, *Spain's Photovoltaic Revolution: The Energy Return on Investment*, SpringerBriefs in Energy, DOI 10.1007/978-1-4419-9437-0_3, © Pedro A. Prieto and Charles A.S. Hall 2013

**Table 3.1** The evolution of solar photovoltaics in Spain (1994–2011)

| Year | Energy sold (GWh) | Installed power (MWn) | Added installed power (MWn) | Rate of increase (%) | No. of PV installations |
|---|---|---|---|---|---|
| 1994 | 1 | 1 | 0 | 0 | 0 |
| 1995 | 1 | 1 | 0 | 0 | 0 |
| 1996 | 1 | 1 | 0 | 0 | 0 |
| 1997 | 1 | 1 | 0 | 0 | 0 |
| 1998 | 1 | 1 | 0 | 0 | 12 |
| 1999 | 1 | 2 | 0 | 3.4 | 16 |
| 2000 | 1 | 2 | 1 | 33.7 | 45 |
| 2001 | 2 | 4 | 2 | 86.6 | 196 |
| 2002 | 5 | 7 | 3 | 89.3 | 795 |
| 2003 | 9 | 11 | 4 | 57.9 | 1,581 |
| 2004 | 18 | 23 | 11 | 99.6 | 3,266 |
| 2005 | 40 | 47 | 25 | 108.1 | 5,391 |
| 2006 | 105 | 146 | 98 | 207.6 | 9,875 |
| 2007 | 484 | 690 | 544 | 372.6 | 20,284 |
| 2008 | 2,528 | 3,398 | 2,708 | 392.4 | 51,310 |
| 2009 | 6,074 | 3,397 | −1 | 0 | 51,312 |
| 2010 | 6,495 | 3,841 | 444 | 8.4 | 55,016 |
| 2011 | 7,398 | 4,246 | 405 | 9.5 | 57,903 |

*Source*: CNE Report, May 2012.http://www.cne.es/cne/Publicaciones?id_nodo=143&accion=1&s oloUltimo=si&sIdCat=10&keyword=&auditoria= (The link slightly changes figures for administrative reasons)

## Electrical Output of Solar Photovoltaic Plants

When the outputs from all installed technologies in Spain are combined, there were, officially, 6,074 GWh generated from 3,397 MWn of installed power in 2009. That is 1,788 MWh generated per year for each nominal installed MWn or roughly 149 twelve hour days' worth at full capacity. The actual electricity delivered to society, key in any EROEI analysis, must be corrected/adjusted/reduced, however, for several factors that are not considered by the Comision Nacional de Energía (CNE) in their measures of electricity generated at the low tension digital meters. We will explore this in more detail in Chap. 5.

## Photovoltaic Technologies Utilized in Spain

The majority of solar installations are fixed (i.e., pointing in one direction), with smaller percentages of one- or two-axis tracking that keep the device pointing toward the sun (Table 3.2). The basic technologies implemented include all the modern PV types: crystalline silicon, HCPV, or thin film. The average size of installations is large, and most are on the ground (Table 3.2).

**Table 3.2** Structure of PV installations in Spain as of July 2009

| | |
|---|---|
| Fixed plants | 63% |
| One axis trackers | 13% |
| Two axis trackers | 24% |
| HCPV | 0.6% |
| Thin film | 2.1% |
| Crystalline silicon | 97.3% |
| <2 MW installations | 36% |
| 2–5 MW installations | 20% |
| >5 MW installations | 44% |
| Rooftop installations | 2.2% |
| On the ground installations | 97.8% |

*Source*: July 2009 report on the status of PV in Spain. Asociación de la Industria Fotovoltaica, ASIF

Spain, in contrast to Germany, installs PV module systems mostly on the ground, where trackers are more suitable than in rooftop installations. The reasons for using trackers can be seen easily when analyzing the generated energy over 1 day with the different systems (Fig. 3.1).

The energy gains with the trackers are especially interesting when there is a decreed upper limit to the instantaneous generation of a given system, as originally was the case for Spain. This was done to encourage the development of smaller PV plants. Here, the higher premium tariffs were distributed only to installations with an upper limit of 100 kWn per individual legally registered installation. Under these circumstances, two-axis tracking systems are capable of capturing 30–40% more energy than fixed systems for the same installed power, without exceeding the 100 kWn output limit at the inverter (Fig. 3.1). This higher generation of the tracking systems needs to be balanced with their higher production and maintenance costs. These costs include the parts and electronics used to run the solar trackers of these systems and the higher consumption of energy to run them. Therefore, if subsidies are high enough, it encourages the installation of trackers.

When the limits for an individual PV plant were increased to 10 MW, in order to encourage the more efficient, larger transformers and inverters, the tracking systems lost their advantage. Consequently, most of the new and large solar PV installations have been fixed installations. Lower module prices for fixed installations and lower subsidies for trackers naturally led to that conclusion in the business plans of PV developers for the last few years (Fig. 3.3).

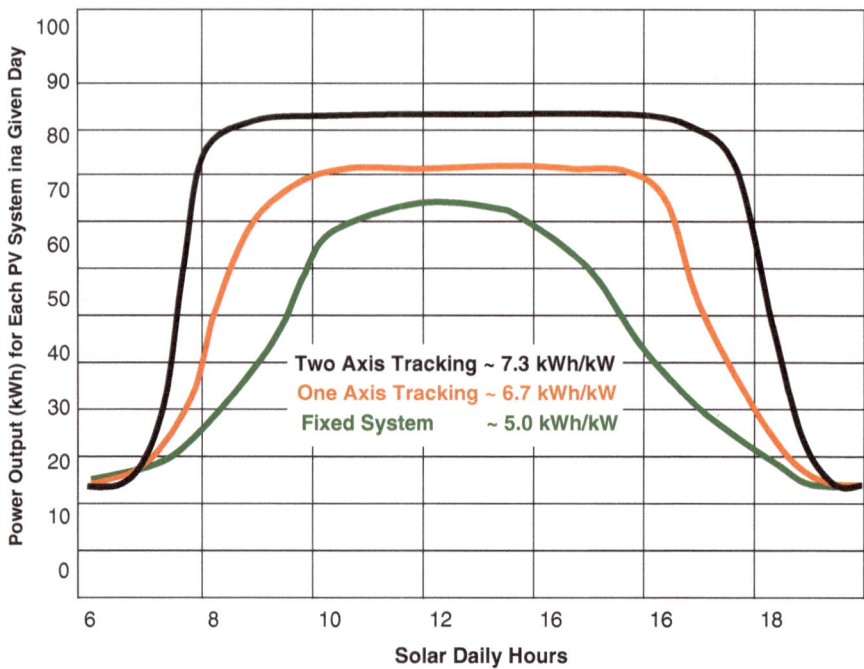

**Fig. 3.1** Typical plotted instant power through time for a fixed solar PV system, an axis tracking PV system, and a two-axis tracking system generated during the day

## Large Government Subsidies and the Solar PV Installation Boom

The Spanish government operated under the philosophy that Spain needed as much domestically produced power as possible. In March of 2004, the government created a premium feed-in-tariff scheme similar to solar PV systems in Germany's through a Royal Decree, which is a law that is signed by King Juan Carlos I (later updated in several successive Royal Decrees). This law allowed solar power operators to charge 5.75 times more than the general electricity rate for the next 25 years and then 4.6 times more than the general electricity rate onward. This charge was applied first to the electric power utility which connected the PV plants to the grid and then to the CNE. The government is obligated to make up for the difference; thus, the consumer indirectly pays for the Spanish premium tariff through taxes.

The government subsidies have been effective, through these "premium tariffs," in promoting on-the-ground installations. They were made available with the idea that this policy would encourage research, development, and innovation, which in time would reduce costs. This is the usual government and business point of view that massive productions always produce cost reductions with the eventual hope that in time, PV electricity would reach so-called grid parity. Grid parity is the time when renewable electricity can be produced at the

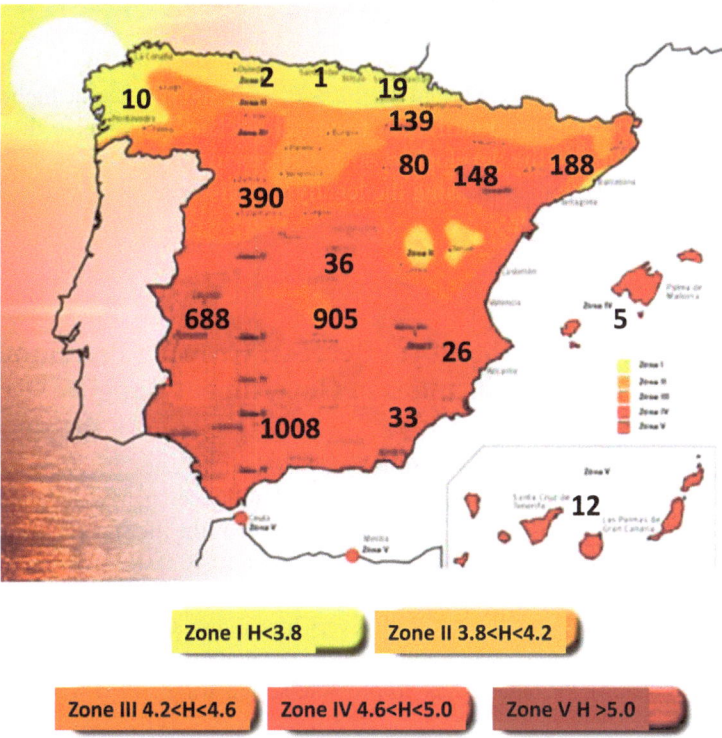

**Fig. 3.2** Daily average irradiance for each climatic zone in Spain (kWh/m²). *Source*: Agencia Nacional de Meteorología. Generated from isolines of annual global solar irradiance on the horizontal surface. The numbers correspond to the installed solar PV power for each autonomous region at the end of 2010 in MW. *Source*: Comision Nacional de Energía

same price as conventional energy sources such as oil, coal, natural gas, nuclear and hydroelectric power. The program was intended to accelerate the technical progress of PVs, so that in time they would be competitive with fossil fuel-derived electricity. The Spanish policy for solar PV alternative sources of energy provided tariffs slightly lower than those in Germany, which seemed to be reasonable due to higher solar irradiation in Spain (Fig. 3.2). The tariffs were also intended to give stability to investors.

Initially, the Spanish FIT targeted installations with an upper limit of 100 kWn (nominal kilowatts) per site in an attempt to encourage many medium-sized Spanish companies to invest. However, soon investors realized the legal possibilities to get more of the tariff with much more total MW by installing many units of 100 kW in their solar parks. The only problem with this scheme was that engineers were forced to have individual transformers and low-tension digital meters in each 100 kW plant because of legal restrictions. This increased the overall costs and reduced the efficiency of PV plants compared to when only one transformer was required for the entire solar farm/plant (Fig. 3.3).

## Bending the Rules

Another approach used to get around the 100 kW limit and still get the highest possible premium or subsidized tariff was to install trackers to optimize the energy captured. This allowed the developers' solar devices to get the maximum rate for more hours in a day without passing the legally established upper limit of 100 kW at the output of the inverter.

## Ups and Downs with the Royal Decrees: Growing Premium Tariffs vs. Growing Difficulties in Government Budgets

The important thing is that all of these very favorable tariff rates were granted for at least the next 25 years. This essentially guaranteed utility companies a premium rate of 575% of the market rate for the next 25 years, which is the life span of their investment. This was an effective rate of nearly half a euro (0.46€) per kWh produced—nearly six times the market price for electricity at the date of entry into the grid. Investors swarmed to take advantage of this very high return. On May 25, 2007, a new regulation was made public, superseding the old premium tariff, in an attempt to stop or at least slow down the industry and its promoters. This was thought necessary when

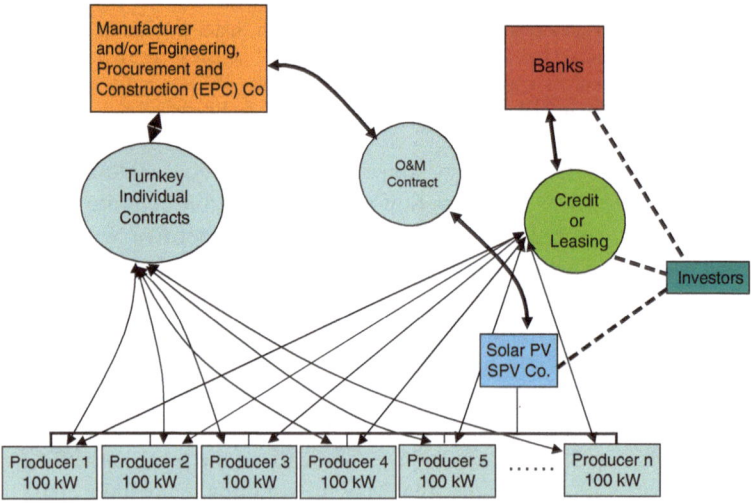

**Fig. 3.3** Financial and legal structure used by Spanish investors to generate more revenue from one project and still meet the legal limits of no larger than 100 kW. They did this by fragmenting their PV projects into several 100 kW separate legal limited liability companies, each owning a plant within a larger park that usually shares some elements and the evacuation power line. Therefore, a huge park can be owned by one big investor while complying with the legal limit of not exceeding the 100 kW power per individual plant

the Spanish government realized that the 2010 production target was already achieved by 2007. The new regulation, simultaneously, tried to protect those who already had PV plants in construction but had not yet connected them to the electric grid (the last requirement to be entitled to the tariff), while at the same time trying to freeze the construction of such super subsidized solar PV plants. The main reason why the government wanted to stop solar PV construction was that the earlier Royal Decree did not foresee the huge number of investors that would be attracted by the favorable tariff rates. At this point, the Cortes Generales (Spanish Congress) wanted to avoid increasing the monetary burden that the feed-in-tariff was creating on the national budget for the next 25 years. At this time, the Spanish economy was booming.

The government realized, although too late, that the margins were much higher than necessary to make this business attractive in a sunny country. Madrid woke up to the fact that the electric system was very much indebted with a protected tariff. The reasons for this subsidy are beyond this book but are related to the moratorium on nuclear energy in Spain. The Spanish government had to come to reality soon, and it began to allow substantial price increases on consumers. The solar PV tariff program was in effect a massive government-endorsed transfer of funds from the public to investors. Recently, solar PV promoters have demonstrated that the tariff deficit in Spain (amounting to 24 billion euros in 2012) was not created from the subsidies for renewable energy but, instead, was due to many other complex factors such as the nuclear moratorium.

Despite the Spanish government's timid attempts to freeze the high tariffs, the subsidized premium tariffs that remained in effect continued being very attractive. This is the main reason for the spectacular expansion of solar PV power installations in Spain in 2008. The highly publicized and admired environmental reasons were far less important as an incentive. Unfortunately, this Royal Decree worked in the opposite direction to the stable market conditions that had been planned, and created a lot of anxiety in the investors that had already planned, and in many cases, committed, some investments but now wanted more. So a frantic race for installation started, and soon the government realized that the second Royal Decree of 2007 was not going to be sufficient to freeze installations and hence limit the long-term commitment of the government to premium tariffs. This provoked a race to have new installations that many developers promised to finish within the time before the existing tariff rates were changed. It also incentivized a frantic search for PV modules, inverters, solar trackers, and 100 kW transformers from wherever they could be found. By midyear of 2008, the government calculated that they would end the year with more than 3,000 MWn installed, far beyond their budgetary plans.

That summer global financial crash of 2008 impacted the Spanish economy seriously. This created panic in Madrid because so many installations had been added that year at such high premium tariff rates. Paying those subsidies would become a larger long-term burden on the government than previously expected. On September 26, 2008, the government hastily passed a new regulation aimed at fixing new lower tariffs, just a few months after the second one, in order to decrease the subsidies from 0.47 to 0.32 and then to 0.27 €/kWh. This affected new PV installations and other installations that were already under permit and legal procedures, with investments already made based on higher tariffs.

As a consequence, the Spanish government set precise limits (quotas) for the total amount of power that could be installed for 2009 and 2010, 500 MW for each year. The new subsidies were to be made available to the solar PV license owners only if the government had registered them. This officially required the owners to register their requests for development of a given plant, and it included the requirement that the owners placed bonds for their given PV plant in order to assure the government that the plant would be built within one year of the initial assignment.

## Theoretical Interest Is Still Alive Despite the Reduction in Premium Tariffs and Assigned Quotas

Another important aspect to consider is that, despite the low number of new PV installations in 2009 and 2010, the number of formal requests to build PV Installations was several orders of magnitude higher in 2009 than the number of available quotas (Ministry of Industry, Tourism and Commerce 2010). In some cases, the list of requests from 2010 would fill all the available slots up to 2014, even though the credit crunch had limited many investment options.

On the other hand, all of the installations of all types throughout 2009 and in the first quarter of 2010 did not even fulfill the solar PV government-assigned quotas. Several factors may explain these discrepancies among opposing forces:

1. The financial crisis that is shocking the whole world, and of course Spain, is operating against more installations. The rate of installation of PV declined sharply everywhere.
2. The sharp decrease of PV module prices would seem to operate in favor of more installations. This decrease is partly attributable to technological improvements with more manufacturers offering new products and partly due to stocks accumulating for lack of sales and excess new production capabilities, where more growth was expected in Spain and globally than actually took place after the 2008 financial crisis.
3. But the huge financial subsidies required to incentivize installations bring us back to the issue of the degree to which modern renewable energies are still underpinned by a fossil-fueled society. The "crisis" in investment for solar probably has been enabled by the end of cheap oil and the global peak of oil production. We will come back to this issue later when we analyze the energy costs of solar PV.
4. Banks have strengthened limits on credits or leasing for these types of projects, which means that due diligence requirements are now much more complex and entrepreneurs are required to provide much more collateral than in the past.
5. Installations of rooftop modules (>20 kW), originally planned for large roofs on malls, commercial centers and industrial premises, with a minimum critical mass for developers, were also involved in the financial crisis. Many of them closed down and the legal security needed for a minimum of 25 years renting contracts was often not clearly reliable for financial entities.

6. Smaller installations on household roofs have no critical mass for big promoters or banks, the main boosters of this sector in Spain, or when thinking of a massive deployment to replace fossil fuels. The administrative procedures seem to be too complex for individuals. The urban structure contains more large private, individually owned dwellings or detached houses which make installations more difficult to optimize due to lack of proper solar orientation in many places and shadowing effects. The legal requirements were complex.

7. The plummeting present prices of solar PV modules, probably because of the huge stocks available at the time, may still make investment with lower premium rates attractive. Whether the investment will continue to be profitable in case of a renaissance of solar PV projects and given the possibility of a sharp increase in the prices of PV modules and the present low subsidy rates is still unknown. If a real push through renewed premium subsidies occurs, it may result in a second sharp increase in PV installations.

8. The possible reassignment of premium tariffs by government creates uncertainties for investors, as the queues of applicants are so large. The tariffs granted by the government are fewer every quarter and published only at the beginning of each quarter, amid the continuous government concerns about the increasing costs of the premium tariffs. Many projects cannot receive approval from banks until the tariff is granted and the module prices fixed with the supplier. These are two big uncertainties for possible investments into any solar PV project.

The net effect of all these rapid changes and uncertainties is a solar future impossible to predict.

## Dancing with the Prices

PV modules are the most important component of any photovoltaic systems. They were subject to market pressures as the industry expanded greatly and then contracted. Prices went up sharply initially in Spain (Fig. 3.4) and subsequently in all related and neighboring countries until the deadlines fixed for the solar feed-in tariff by the Spanish government in September of 2008 expired.

## Impacts on Spain's Industries

Electric rates went up some 16–20% in 2008. Consequently, many electricity-intensive industries with low profit margins (tiles, ceramics, agriculture with water desalination or pumping needs, steel, aluminum, and other metal industries) were caught between the global downturn in demand and the increases in electricity prices. Many threatened the government with leaving the country to other places where electricity is still heavily subsidized. This also shows clearly the sensitivity of many Spanish industries, including PV manufacturers, to electricity prices.

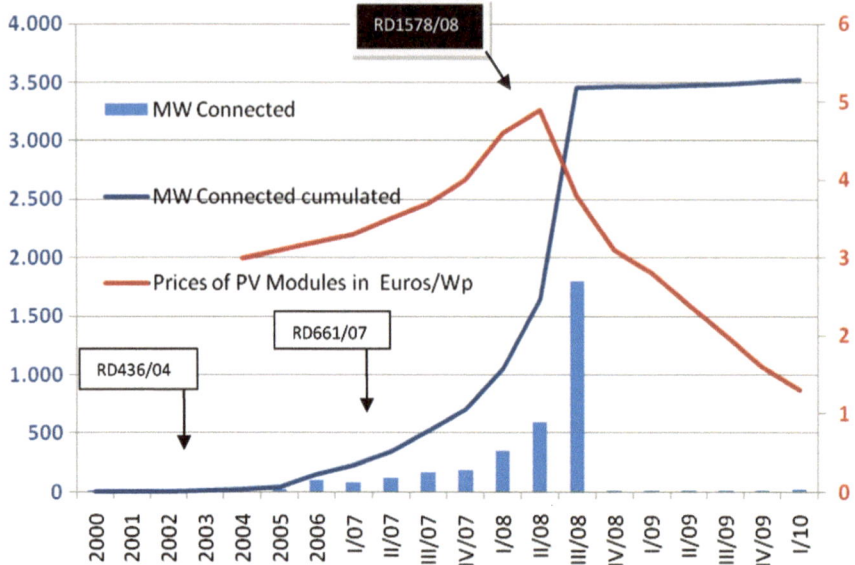

**Fig. 3.4**  Solar PV installations connected to the grid (*blue*) and price changes of PV modules (*red*) from 2000 to the first quarter of 2010. RD stands for Royal Decree

That made the Spanish solar PV industry and all the auxiliary sectors associated with it fade once again as the considerably reduced margins for investors went from an economic payback of 12–15% to 6–8%. These profit reductions occurred because the large investments in new capital equipment that the industry made based on the government-inspired projections of high permanent production rates were never going to be amortized with the new government plans. Many small- and medium-size solar PV companies filed for bankrupcy, and many others started to experience huge financial problems. BP closed two Spanish factories. Many intermediaries, importing from China, found themselves with their stores full of modules but no customers while still having to pay interest and principal to their creditors.

Since 2000, Spain has experienced a number of different new government regulations, which have caused the private solar industry to blame the government for constantly changing the rules of the game. On the other hand, the government has blamed promoters and manufacturers and, especially, investment firms soaring like vultures over what they called "solar investment funds," all of them looking for-short term high profits and margins, rather than for long-term research and development efforts. But the market is the market and this government never questioned the "free market" even as it changed the rules upon which it operated. Banks and investment firms were offering "special funds" granting 12% of equity, or higher, to their preferential customers without mentioning that 4/5 of their incomes were provided by the government in the form of subsidies through premium tariffs, which of course were derived principally from the general activity of the Spanish economy which is run mostly on fossil fuels.

It seems that if the fossil fuel society does not finance and subsidize these so-called renewable energies, they will not be able to develop by themselves to a mature state in which these technologies will be able to service the modern society at the present levels of consumption while they breed themselves, and as we see here, it is very difficult to carry that out on a large scale.

## ...And Spain Died of Success

After a couple of years of success hype, the industry itself is starting to acknowledge that the rate of growth was not something they could easily swallow. In other words, the previous growth rate of the solar industry was unsustainable partially by recognizing that the Spanish government could not go much farther in subsidizing the industry at those levels.

Figure 3.4 shows how the consecutive and frantically passed laws were instrumental in first promoting, and then freezing over a very few years the development of solar PV installations in Spain. The ramp up of solar PV production as the incentive laws were passed was subject to some inertia but the freezing was much faster. Since 2008, different Royal Decrees and dispositions have worsened the position of the developers that invested in solar PV plants and generation. Among them, the most important Royal Decrees are:

- RD1003/2010 August 5th. To regulate the legal requisites, the so-called decree against fraud.
- RD1565/2010 November 23rd. To preassign new developers requesting new PV installations.
- RD Law 14/2010 December 23rd. Limiting the income of solar PV plants to a certain maximum number of sun*hours peak in a year, as interpreted from a previous RD Annex and depending on the type of technology and area of irradiance in Spain. This was limited, in principle, to 3 years, in exchange for a 3-year extension of the 25 years grant for premium tariffs. Many promoters and the solar PV industry have claimed losses of revenue for these 3 years of up to 30%, which in a certain way correlates with the overcapacity installed in these plants (watts peak versus the officially registered watts nominal at the output of the inverters).

## Difficulties Associated with the Success of the Spanish Solar PV Program

In some respects the solar program was remarkably successful. By 2008 Spain had installed about 3,500 MWn, more than 800% of its initial target of 400 MWn (Table 3.1). The largest installation occurred in 2008 with 2,759 MW installed, almost half of the world's PV installations for that year. Even though Spain's installations were more than in any other country, it represented a cumulated production capacity of only some 3% of Spanish electricity production and about 1% of its overall primary energy use.

Although Spain achieved a great deal of physical success with its PV program, there were many associated problems and costs that led to its financial failure. By 2007 and at the beginning of 2008, many more installations had received instrumented credits from the banks (in those days easily granting a leverage of 80% on the total project), purchased the equipment at whatever the costs, received the construction permits from regional authorities, and had embarked on the developments, although they were often still not connected to the grid. They were given the credits in the hope of connecting to the grid before the deadline of September 2008 and therefore getting the best possible premium tariffs/subsidies. In those days of still wine and roses, the government did not dare to tell an industry that was installing thousands of MW simultaneously: "ladies and gentlemen, MW number 400 was just connected to the grid and granted the highest premium tariff, but the rest under installation will have to go back to the banks and tell them that while you expected to receive 46 cents of euro/kWh for the next 25 years (plus an inflation correction), you will be receiving much less."

## The Increasing Use of Imported PV Components and Equipment

Although the initial goal was to use mainly Spanish-produced solar components, the industry was growing so rapidly (triple digits in 2007 and 2008) that local producers did not have enough production capacity to produce the solar modules to supply the Spanish market. In addition, the attempt to meet government goals pushed the Spanish silicon wafer industry so much that they had to raise prices considerably to cover new investment needs for new production facilities, as present ones were clearly not sufficient. According to some reports from the Ministry of Industry, some 2.8 billion euros in modules were imported in 2008 at prices of 3–5 euros/Wp. Spanish-made modules had to compete in price; therefore, the only advantages with respect to those imported from outside the EU were the import duties, which for such a liberalized country were not enough of a barrier to prevent the entry of the first shipments of imported Chinese modules. That was not important for many developers, as they were much more concerned about the imminent deadline to get the premium tariff than the potential lack of quality, especially of the first imports. The time pressure caused some developers and importers, in some emergency cases, to ship by air to comply with the September 2008 deadline imposed by the government.

The situation became much worse in that by 2010, China had captured a significant part of the global solar PV market because it could manufacture modules much more cheaply than elsewhere. Apparently, this was due to cheap labor and a strong governmental commitment with long-term policies that allow China to foresee and meet world market demands and also to China's attitude to weather market fluctuations that can kill businesses in other countries.

Today, many American, Japanese and European solar PV producers have a joint venture with a Chinese manufacturer and are producing modules with European, American, or Japanese brand names but made in China. Thus, they could be in full compliance (theoretically) with all the hidden barriers that Americans, Europeans, and Japanese erected (e.g., standards, ISOs) to preserve their own "original equipment" manufacturers. The companies were in fact collaborating with the Chinese to corner the market and to produce cheaper modules that appeared to be made in Western countries, or were somehow endorsed by Western companies. This was necessary for them to compete and survive in the short-term, but it will backfire against them in the long-term.

The Chinese government's commitment includes explicit policies of massively producing or purchasing essential raw materials (silicon ingots, wafers and cells, aluminum, steel, copper, tempered glass, etc.) and in some cases "cornering the market," which, of course, gives them a very competitive edge. Their focus is clearly to use their enormous purchasing capacities to hoard materials and then sell their products later to the world to obtain a manufacturing advantage. The Chinese are very disciplined in this. Their ability to maintain their policies because of strong governmental assistance, while the rest of the world crumbles (while parroting the virtues of the free market) is considerable. Their long-term views are much longer than the present 2-year strategic plans in Western companies or the market faith of much of the Western world. This has led to a very stable PV business environment in which China can increasingly outcompete the rest of the world and their reliance on market mechanisms. One example is the internal purchasing capacity in 2009 and 2010, when the Chinese government absorbed internally a portion of its factory throughput in order to keep their huge investments in manufacturing facilities working even in the middle of a world financial crisis.

An additional reason is that China's labor and working conditions are the most "competitive," in the classic business as usual (BAU) terms. This is true because Chinese labor and factories have very little embodied monetary and energy costs compared with the costs of American, European, or Japanese equivalent workers (i.e., safe and comfortable working conditions and general worker education outside their specialty, wages and benefits are generally at a much lower level than in OECD countries). Most European, US, and Japanese producers have opted to survive by transferring their "maquilas" (factories) (i.e., low-cost labor assembly plants) to China, while at the same time attempting to keep the royalties from their R+D and new developments. But China has a fast learning curve in these issues too (and many excellent engineers) and is already producing products using local designs that comply with all stringent European standards and quality inspections.

Nevertheless, China's solar PV industry has also been affected by the world's economic and financial crisis, since they were counting on supplying an exponential expansion of PV installations worldwide. There are signs that China's solar PV industry is restructuring as the domestic market attempts to swallow the slowdown in foreign orders.

## Solar PV as a Financial Product, Rather Than a Renewable Asset

The Spanish government consecutively passed laws that had unintended negative consequences. They generated grants in a form similar to state bonds at fixed, high interest rates over an unlimited period, with a minimum guaranteed return over 25 years. The real world, that fossil-fuelled world with energy surplus in the form of surplus wealth available as financial investment packages, immediately saw this as a very secure "financial product," rather than an opportunity to improve an incipient technology. As a consequence, a great race ensued for the acquisition of rights to install PV plants within a government subsidized business environment. In 2007, the government realized that because their incentives were far larger than originally planned, the 2010 target was going to be fulfilled soon and that much more capacity was on the way. But it was already too late.

In March of 2009, the General Director of energy policy in the Ministry of Industry held a conference in Palma de Mallorca. Representatives of the PV industry were extremely confused and concerned by the freezing of the feed-in-tariff, by the limits imposed to the yearly installed base, the lack of horizons beyond 2011, and the continuous changes in the legal conditions with every new Royal Decree passed every few months. This created great legal insecurities for manufacturers and investors, just after they had invested heavily in manufacturing plants, licenses, and R&D facilities. They had planned for a long-term continuance of the program.

The General Director responded that the costs in premium tariffs represented some 7% of the cost of the whole national electric bill for every additional 1% of solar PV electricity, and that the government could not continue this unsustainable subsidy. Others pointed out that electric bills increased by some 4% for each 1% of additional solar PV electricity. Some environmentalists protested, arguing that any "investment" (meaning subsidies, premium tariffs, and so on) in "renewable energies" would be worth it in terms of the environmental benefits from not burning fossil fuels—without understanding that the subsidies to solar electricity were subsidies of not only money but also the fossil energy that the money represented.

In reality, what happened was that the very large flows of money basically went to the already installed solar farms, i.e., those already entitled as *solar PV financial products*, rather than to research, development and innovation (R&D&I), as it is known in Spain, of new solar technologies.

Several first-class banks started offering their customers what they called attractive "financial products" (in fact, the solar farms, which were an interesting "product" because they had hidden subsidies from all Spanish citizens) with some 12–15% *financial return on investment* (FROI) and a much higher *return on capital equity* (ROCE), considering that the out-of-pocket capital required was only 20% of the total turnkey price and 80% was going to be financed by banks. A flock of other banks and financial investors immediately came to the trough.

This high return existed when considering that turnkey prices of the solar PV plants in those days were on average 5 €/Wp for fixed plants and up to 7 or 8 €/Wp for two-axis tracking plants, mainly due to the high prices of the premium subsidized electricity produced. This meant that the subsidies were some five times the value of the electricity produced, especially for those PV plants originally covered by the March 1, 2004 Royal Decree. This is how modern renewable energies, supposedly born to support a sustainable world, became one jewel of the most unsustainable of human activities, financial greed.

Greed also affected some foreign manufacturers who took advantage of this business more as a profitable short-term investment to exploit a flourishing market, by treating it more as a profitable business of producing and selling more and more solar modules, rather than having a commitment to progress in R&D. They invested heavily in additional manufacturing equipment and factories, to cover the demand of what they sought would be an inexhaustible market, following the trends of Japan, Germany, and now Spain.

# Fraud

The incredible race to reach the time limits of the 47 euro cents per kWh subsidies (September 2008) led to a number of fraudulent activities which new regulations are still trying to untangle and fix. A significant proportion of the PV modules said to be installed before the deadline of September 2008 were in fact not connected to the grid by that date and some engineers, public officers, electric companies' employees and others signed preliminary documents certifying, somehow, that these plants had been connected.

When these engineers found that the government was going to audit if in fact the meters had been running before September 2008, many installed just a few modules (which in those days were very scarce) per each of the individual 100 kW plants just to make the meter run. The scandal was so large that the government sent a number of inspectors to tour all of Spain and to impose heavy fines for those caught engaged in fraud. Even mayors of municipalities gave licenses without the precise requirements having been met. The General Directorate for Energy Policy and Mines passed a resolution on October 23, 2008, to inspect solar farms. The Comisión Nacional de Energía (CNE) released a report after inspecting 294 installations in 30 different solar farms with a total combined power of over 27 MW. The conclusions were that only about half (13 installations amounting to 12.5 MWn) were legal, with all the proper documentation and generating prior to the deadline. The rest of the PV installations presented anomalies in documentation, could not prove they were generating energy before the deadline, or did not even allow inspectors to audit them.

However, the government finally recognized its impotence to inspect about 2.5 GW of scattered plants throughout Spain due to a lack of sufficient inspectors. Even this report proposed severe sanctions to those infringing regulations. Another report from the CNE gives a good idea of the importance and seriousness taken by the Spanish Administration to crush the frauds and the difficulties created by the policies for renewable energies. Some promoters who were short of PV modules but realizing that some of the potential inspections will take into consideration the date in which the sealed digital meters started to run with the generated electricity, resolved to install a few modules in each 100 kW unitary installation, to make the meter run. Presently, inspectors also double-check these attempts to fudge the inspections by evaluating the generated energy in relation to the days of the year and theoretical installed power. The authorities are asking for custom documents proving that modules and other equipment were in place before the due date and also requesting delivery bills at the sites. This was done under the threat of expelling the proven offenders from the tariffs, after inspection, making them use the open market rather than subsidized prices and asking them to refund all the invoiced excess from that time to the date of inspection. Alternatively, they offered possible offenders a sort of amnesty to those voluntarily downgrading from the 46 to 32 cents kWh by October 2010.

In 2010, a second type of fraud was also uncovered and released to the press by sources within the Ministry of Industry. It identified a few unidentified PV plants that were illegally and temporarily installing diesel generators and connecting them to the bus bars to produce subsidized energy at premium tariffs, as if it were solar PV-generated energy. It is, however, unclear whether this type of fraud is significant or, most likely, has been one or a few isolated cases. Interestingly, this case was highlighted by sources close to ministerial deep throats, amid a campaign to discredit solar PV plants, which is being accused these days of costing about 2.5 billion euros/year in subsidies, while covering only 2.1% of the national electricity demand. The PV industry answered by saying that this was part of a discrediting campaign and that this type of fraud has only occurred in <0.1% of the installations. In comparison, wind energy subsidies cost the government much less while generating about 15% of the national electricity demand.

The PV Industry Association and the Association of Solar PV Producers have angrily asked the government to identify the cheaters immediately and to punish them with heavy fines and to cancel their generation permits. In fact, for CNE and the electric power utilities measuring the digital sealed meters, it is very easy to identify any type of fraud like this. Solar PV plants are required by law to have a memory of the production of each quarter hour, stored for as long as required, usually per month. With these stored data and the precise meteorological data per region and day, inspectors could easily discover who was cheating.

## The Financial Support to Renewable Energies from a Fossil-Fuelled Society

As the government had stated it, the present annual budget in subsidies (premium tariffs) for 3,547 MW of solar PV plants represents some 2.33 billion euros/year for Spanish taxpayers. Amidst the global financial crisis, this has forced the government to enter into a deficit without precedents. But money has value only if there is energy behind it to produce, transport and market the goods and services. If the subsidies are coming from the taxpayers, i.e., the society in general, which is fuelled almost entirely by fossil fuels, a monetary subsidy is the equivalent of a fossil fuel subsidy. All of these issues lead into the question: "what is the net energy delivered by the Spanish PV revolution?"

In conclusion, the main problems were that:

1. The program bankrupted its own funds due to very high costs for the subsidies thought necessary to lure contractors into constructing what were otherwise essentially high-cost relatively low yielding (or much lower than originally expected) investments.
2. The growing need to use increasingly imported components, due to too rapid a growth and associated globalized and open market issues. These distorted the initial goodwill of a national development plan for the creation of the solar industry and job creation and minimized the R&D efforts.
3. The solar PV venture was increasingly viewed as a financial, rather than an energy venture, and this approach led to financial failure.

However, we believe that the major problem with the solar program was that solar energy is still a relatively-low energy-yielding, i.e., low EROI technology. When all of its energy costs are included, and with the understanding that the energy investment comes mostly from fossil fuels in large and complex ways, the resulting low EROI leads to the necessity of excessive financial subsidies that undermined the financial success of the program. This makes the desired objective to reach grid parity very difficult despite the important goal, which we agree with: to make as much of our energy from renewable sources of energy as makes sense. What makes sense, however, is not quite clear.

# Chapter 4
# Calculating the Energy Return on Energy Invested (EROEI or EROI) for Spain's Solar Photovoltaic Energy

Neither an organism in nature nor human society can live without energy. Gaining that energy, whether it is solar, food, wood, coal, oil, natural gas, hydropower, wind, solar PV, or anything else, requires specific systems devoted to obtaining that energy. These systems in turn require both financial and energy investments, both initially for their construction and over time for their maintenance and repair. The energy returned by a system able to deliver energy to society (called ER or $E_{out}$) divided by the energy input required to get it (EI or $E_{in}$) is called the energy return on investment (EROI or sometimes EROEI, with the second E used to refer to the use of energy in the denominator). The term is related to "net energy" (e.g., White 1949; Odum 1973; Hall et al. 1979) and some aspects of "life cycle analysis" but is more explicit. It was derived originally from thinking about fish migration as an energy investment (Hall 1972), and its first use for societal energy was in Hall et al. (1979) and Cleveland et al. (1984). Usually this relation is derived either as a year by year average (typical for a nation's oil production) or from costs and gains over the life span of a particular or typical individual well or unit. The latter depends upon a key variable: the expected duration time, or so-called life cycle, during which it is producing energy, either theoretically or, ideally, measured. In photovoltaic systems, this period is usually considered 25 years from the point of installation. We explore this in detail in the following chapters.

The energy return on investment (EROI, sometimes EROEI) is expressed as a simple ratio of energy returned from an energy-gathering activity compared to the energy invested in that process:

$$\text{EROI} = \frac{\text{ER}}{\text{EI}} \quad \text{or} \quad \text{EROI} = \frac{\text{energy returned to society}}{\text{energy invested to get that energy}}$$

The numerator and the denominator are necessarily assessed in the same units so that the ratio so derived is dimensionless, for example, 30:1, which can be expressed as "30 to 1." This means that a particular process yields 30 J on an investment of 1 J (or kcal per kcal or barrels per barrel).

P.A. Prieto and C.A.S. Hall, *Spain's Photovoltaic Revolution: The Energy Return on Investment*, SpringerBriefs in Energy, DOI 10.1007/978-1-4419-9437-0_4, © Pedro A. Prieto and Charles A.S. Hall 2013

The output is normally expressed as, e.g., 20:1, for twenty units returned for one unit invested, as is the approximate value for North Sea petroleum at this time. While this equation is very simple, the devil is in the details. Major sources of uncertainty include availability of trustworthy numbers, boundaries of the analysis, assumptions about the quality of different energy types, and ways to deal with real inputs for which there is only financial, but not explicit energy, values. All of these factors may vary considerably during the actual operation of the system compared to the theoretical or laboratory values.

In the case of photovoltaic systems, ER is composed of the nameplate energy output of the actual PV collector ($E_{nameplate}$), which is the output energy that the device is rated at and is theoretically capable of producing, multiplied by a series of factors varying from zero to one that we call *loss* factors. These loss factors decrease the theoretically generated energy from the nameplate output, i.e., $ER = E \times l_1 \times l_2 \times l_3 \times \cdots \times l_n$. ER is the energy delivered to the grid or society, i.e., the existing energy-delivery infrastructure. What these variables are and their values are estimated in Chap. 5 for Spain.

EI, the energy invested in order to produce the energy output, is composed of a series of additive investment factors: $EI = a_1 + a_2 + a_3 + \cdots + a_n$. In general, the more factors included in the denominator, the lower the EROI. The analysis is critically dependent upon the boundaries chosen. Murphy et al. (2011) derived a protocol for assessing the boundaries for an EROI analysis. They sought a balance between the need for a standardized approach, so that different technologies can be compared, and a flexible approach reflecting the needs of particular technologies, questions, and investigators. They recommend that a standard approach be used ($EROI_{stnd}$) and that once this is done, the investigator may change the energy inputs and outputs as they see fit, as long as they are developed explicitly.

In our case, the objective is to get as comprehensive an analysis of EROI as possible for a real operating PV system in Spain, so although we do calculate the $EROI_{stnd}$, we also expand the boundaries to include a very comprehensive list of energy inputs, all of which we believe are provided (mostly) by the use of fossil fuels and all of which are necessary to make the actual PV system work. We recognize that many of these issues are controversial; this is why we attempt to provide a basic sensitivity analysis so that the reader may choose which final numbers to choose. Nevertheless, we believe that our estimate is comprehensive yet still conservative, so that the real EROI is likely to be less than our calculation.

## Energy Outputs (Numerator)

For some energy sources, such as petroleum and natural gas from a traditional oil field, it is easy to calculate the energy return (ER or $E_{out}$). But when the energy, in this case again oil and gas, comes from mixed primary sources (i.e., different types of oil fields) and it reaches society only after complex processes of extraction, transportation, refining, etc., calculating the ER can be a difficult task. Often national mean values are available and subject to internal quality control (see e.g., Guilford et al. 2011; Grandell et al. 2011).

In the case of the Spanish solar photovoltaic systems, specifically those intended for "feed-in" systems (i.e., to be fed directly into the existing national grid), the ER assessment is relatively easy since the energy source is only the photovoltaic module and its accessories and the output energy is delivered directly as electricity to society. The ER is relatively easy to determine, especially in the case of Spain, where all the feed-in plants are invoicing to a national public entity (CNE), either directly or indirectly. In Spain, each of the more than 57,900 PV plants installed is fully accounted for with complete yearly records of operation. Thus, "all" that is needed is to correct the output for the loss factors, which we do in Chap. 5.

## Energy Inputs (Denominator)

Calculating the energy invested (EI or $E_{in}$) for the Spanish photovoltaic system is more complex than for outputs and depends a great deal on the boundaries chosen for the analysis. Other studies have estimated inputs in terms of energy used for only a few of the real energy inputs. In our EROI analysis, we have a very comprehensive list of estimates for the money used per GWn for essentially all of the energy inputs needed. These inputs require conversion factors, called *energy intensities* (i.e., MJ per euro), to calculate their energy costs. These procedures are actually rather well developed but usually lack the detailed information we need in recent years to get good estimates for PV systems. We develop these procedures in more detail for deriving energy intensities, along with some uncertainty analysis, and then apply them in Chap. 6.

## Deriving Energy Intensities for Money Spent on Inputs

Where possible it is desirable to measure the energy used directly (e.g., on site or in centralized manufacturing facilities) in terms of physical units. In fact, there is considerable information about how much energy it takes to make, e.g., a metric ton of steel (about 21.3 GJ) or concrete (about 5.1 GJ) but less for e.g., transformers or instruments or financial services. When we are lucky, there are estimates of the total amount of energy going into a solar PV factory to produce modules and how many units (e.g., square meters of PV devices coming out). These approaches have been used well by, e.g., Fthenakis/Hyung-Chul Kim (2005), Fthenakis/Alsema (2006), Fthenakis/Kim/Held/Raugei/Krones (2009). See also a collection of studies in Bankier and Gale at http://energybulletin.net/node/17219 and Raugei and Fthenakis and Kim (2009) to estimate the energy used to make solar collectors.

While it is usually much more desirable to measure energy use directly, this is often not possible because neither governments nor, especially, businesses have any particular reason to maintain records on energy use in physical units. They focus instead on economic information since they are usually concerned about the bottom line. We start with the truism that there is energy associated with any motion and economic production, which is essentially a series of movements, concentrations, and transformations of the raw materials from nature. Thus, money spent is generally correlated with energy use (with the exception of the purchase of energy itself), as shown in a number of papers from the Energy Research Group at the University of Illinois several decades ago (e.g., Bullard et al. 1975). The similarity in energy used per dollar spent (or dollar of GDP) is especially the case for "final demand," that is, for the goods and services purchased as final rather than intermediate products (i.e., houses rather than intermediate materials such as lumber, steel or concrete). In the 1970s, we had good numbers on the energy costs of essentially all economic products because the Illinois group (Bruce Hannon, Clark Bullard, and Robert Hannon) had examined in great detail the "interdependency" of the US economy, that is, how much each sector purchased from other sectors. This is sometimes called input-output (I-O) analysis. In other words, a final demand item such as a house or an automobile or even an oil well will include energy-intensive-per-dollar raw materials such as cement and steel, as well as less energy-intensive business services. Unfortunately, it is not possible to make this statement now because the work at the University of Illinois was discontinued and there has not been an adequate replacement since (for a partial exception see Carnegie Mellon 2012).

So, whenever it has been possible and measurable, we have stuck exclusively to the physical units of energy. In the rest of the cases, we start with the costs in monetary units and then multiply those costs by their "energy intensities," that is, the energy use associated with the production of one monetary unit of whatever goods or services we might wish to analyze. The fundamental question then is how much energy is associated with different economic activities?

Although some of the present experts in LCAs and EPBTs are very familiar with the use of such energy to economic ratios, which have been used to assess economic activities from Odum (1973) to the more recent ones (i.e., Murphy et al. 2011), they are sometimes critical of our "extended EROI" analysis, which includes wider boundaries than usual and requires the use of many energy estimates based on monetary data. We believe that one can derive a much more realistic EROI by including all of the energy used to create a PV facility even when using quite approximate monetary-derived energy numbers rather than not including them at all, i.e., by not assessing indirect or hidden (but very real) energy costs, without which the PV plants could have not been manufactured, transported, installed, legalized, inscribed, connected, protected, commissioned, maintained, etc.

The most general and defensible, but most approximate, estimate of energy used per monetary unit is the national mean energy used per unit of economic production. This can be derived quite straightforwardly for Spain, for example, by dividing the total energy used in the economy by the GDP. In fact, the Ministry of Industry, Tourism and Commerce in Spain publishes these figures each year as the ratio of tons oil equivalents (Toe)/million euros GDP. For 2008, the year that much of the inputs to the PV plants in operation in 2009 were made, the ratio was 170.94 Toes/million euro. Given that 1 Toe = 41.87 GJ, this represents 7.16 MJ/euro, or 1.99 kWh used per euro GDP generated or consumed.

This is roughly two thirds of the energy intensity (i.e., energy used per dollar) that is found as a mean in the USA—about 8 MJ/dollar (or 11.5 MJ/euro) in 2008, implying that the Spanish economy is somewhat more energy efficient than is the economy of the United States. Given that much of the materials used to create a PV plant were made in other parts of the world, it is interesting to contemplate what this ratio is for the world as a whole. This ratio can be obtained by dividing the world energy use by the world GDP, which gives about 7.23 MJ per 2008 dollar, or roughly 5 MJ/euro. It would seem that the energy intensity for Spain is roughly a third more than the world as a whole. Thus, it seems not to matter hugely whether the input to a PV plant comes from Spain or the rest of the world.

We next ask whether it matters a great deal whether the item came from a more energy-intensive material production industry or a presumably less energy-intensive (per dollar) financial or related sector. Unfortunately, the answer to that question is not easy for Spain since we could not gain access to a comprehensive I–O energy analysis, and it has been many years since the University of Illinois group undertook their research. What we can do is to examine the relative numbers in the United States for 2002 using the Carnegie Mellon green energy calculator (Carnegie Mellon 2012). This provides a means of examining the energy intensity for various goods and services based on their 2002 I–O data, and allows us to get a general idea of the energy intensity of various inputs to the solar PV system, so we can compare them to the mean for the nation.

In 2002, the GDP of the United States was 10.643 trillion dollars. Its energy use was about 103 ExaJoules, so that 9.68 MJ were used per average dollar in the economy to generate the entire suite of goods and services produced. So, the energy intensity for the economy as a whole was 9.68 MJ/dollar. During that year, the energy intensity of various raw materials was 10–20 MJ/dollar. For manufactured goods, it ranged from 4–25 MJ/dollar, on average perhaps twice the national mean (Table 4.1). The energy intensity for financial and business services was about 1–6 MJ/dollar, about one quarter to one half of the national mean. Similar analysis presented in Murphy et al. (2001) suggested that manufactured/engineered goods were roughly 1.7 times as energy-intensive as the national mean. We extrapolate the much more data-intensive proportional energy intensities to Spain by assuming that manufactured or engineered inputs used twice the energy intensity (14.3 MJ/euro) as the national mean (7.16 MJ/euro) and that business and financial services used one third that rate (2.39 MJ/euro). These numbers are then used in Chap. 6 to calculate the energy inputs. We acknowledge that these numbers are very rough. But it is probably not possible to do a much better job until someone does a really good I–O energy analysis for the Spanish economy.

## Calculating an Extremely Rough EROI for a 1 GW Solar Photovoltaic Plant in Spain

Before we undertake a more comprehensive EROI analysis, we first give an example as to how the total monetary costs used to construct and run a PV plant can be used to generate a very rough idea as to how much energy was used to make and run

**Table 4.1** Energy intensity of various sectors of the US economy in 2002 according to the Carnegie Mellon energy calculator

| Category | Sector | | Energy intensity (MJ/$) |
| | Number | Name | |
|---|---|---|---|
| Raw materials | 331411 | Smelting and refining of copper | 21.4 |
| | 212100 | Coal mining | 15.4 |
| | 213111 | Drilling oil and gas well | 11.4 |
| Manufacturing | 331200 | Iron pipe and tube manufacturing | 25.2 |
| | 33131B | Aluminum product manufacturing | 24.3 |
| | 331420 | Copper rolling, drawing, extruding, alloying | 15.0 |
| | 33299C | Other fabricated metal manufacturing | 12.5 |
| | 33999A | Miscellaneous manufacturing | 9.51 |
| | 322500 | Hardware manufacturing | 9.24 |
| | 33329A | Industrial machinery manufacturing | 9.0 |
| | 230301 | Nonresidential bldg. maintenance/ repair | 8.69 |
| | 334413 | Semiconductor and related device | 7.56 |
| | 333295 | Semiconductor machinery | 7.17 |
| | 335313 | Switchgear and switchboard apparatus | 6.44 |
| | 334111 | Electronic computer manufacturing | 4.28 |
| Financial services | 541800 | Advertising and related services | 4.12 |
| | 561400 | Business support services | 2.90 |
| | 541300 | Architectural and engineering services | 2.70 |
| | 523000 | Securities, commodity contracts, investments | 1.53 |
| | 541100 | Legal services | 1.52 |

These values can be compared with the US average for that year of 9.68 MJ/$

the PV system. This allows one to calculate a very approximate EROI. The energy costs estimated this way may be considered low because many of the inputs have higher energy intensities, and they may be considered high because it uses all the money inputs, not all of which (perhaps) should be attributed to energy costs, and because we may overestimate the energy used by financial agencies. But, as we said, it is a first estimate.

A one MW solar PV power plant in Spain will cost about 5.5 million euros to build and another roughly 30% for operation and maintenance over its nominal lifetime of 25 years, for a total of 7.15 million euros. At the mean energy intensity value for Spain of about 7.16 MJ/euro or 1.99 kWh/euro, this PV plant would require 51.2 TeraJoules of energy investments.

This plant will generate 1,375 MWh/MWp net and averaged energy per year as we state in page 49, figure 5.1. In its 25 year lifetime it will generate 34,375 MWh or 34.375 GWh. Considering 1 GWh = 3,600 GJ it will generate 123,750 GJ or 123.75 TeraJoules. Thus, we have as a very first approximation of a comprehensive EROI 123.75 TJ/51.2 TJ = 2.41 units of energy returned per unit invested (2.41:1). This, of course, is a very rough estimate, so we shortly undertake a much more careful analysis in the following chapters.

This might be considered a low estimate of EROI since the energy intensity of the indirect inputs might be on average higher (or lower) than the mean for society used here. If we assume that all the energy inputs are for engineered or constructed items (e.g., things made of steel or cement or whatever), we might assume for our rough estimate that the energy costs would be two times higher and the EROI accordingly only about 1.2:1. Or we might assume that all inputs are for financial and related services, which might use energy say at half the rate per monetary unit as the national mean. Thus, if we assume all indirect costs are for things such as financial services then the EROI might be twice as high, or 6.9:1. But neither of these ratios is reasonable, as the indirect inputs are a large diversity of all economic activities in the Spanish economy. Nevertheless, all this implies that using a national mean is not a bad start for estimating this rough EROI. Later, we compare our results from a more detailed analysis to our EROI value here of 2.4:1.

## Spanish Data Sources: A Good Start with Good Institutions and Public Companies

The renewable energy program in Spain is very well known and understood both by the Spanish public and internationally. Our principal data source for this EROI analysis is the report released quarterly by the regulatory body of the energy systems in Spain, the Comisión Nacional de Energía (CNE) dated June 2010. We use data for the years 2009 and 2010. This includes the energy generated under the "special regime" with premium feed-in-tariff schemes, basically an indirect form of subsidies to renewable energy pioneers, i.e., to those companies that are initiating new technologies. This report includes considerable data on all such systems connected to the grid.

To the best of our knowledge, this is the best public compilation on renewable energies in the world to date because of its completeness. It includes sections on:

- The evolution of the size and nature of renewable systems through time.
- The degree of target compliance with installation and production targets as per the programs in force.
- The remuneration to producers for their energy production by type of energy, energy sold (GWh), number of installations, feed-in-tariffs, reactive energy, voltage sags and swells, etc.
- The annual GWh generated per year and per MW installed for each type of energy.
- The contribution to the total national consumption (which they call "demand").
- The production for Spanish autonomous communities (i.e., autonomous region) and other useful data. This web page is available only in Spanish.

We use these data as the basis for our EROI analysis of actual energy production with PV systems. We also used the reliable and interesting public domain site of Red Eléctrica Española (REE). Red Eléctrica of Spain was the first company in the world dedicated exclusively to power transmission and the operation of electrical

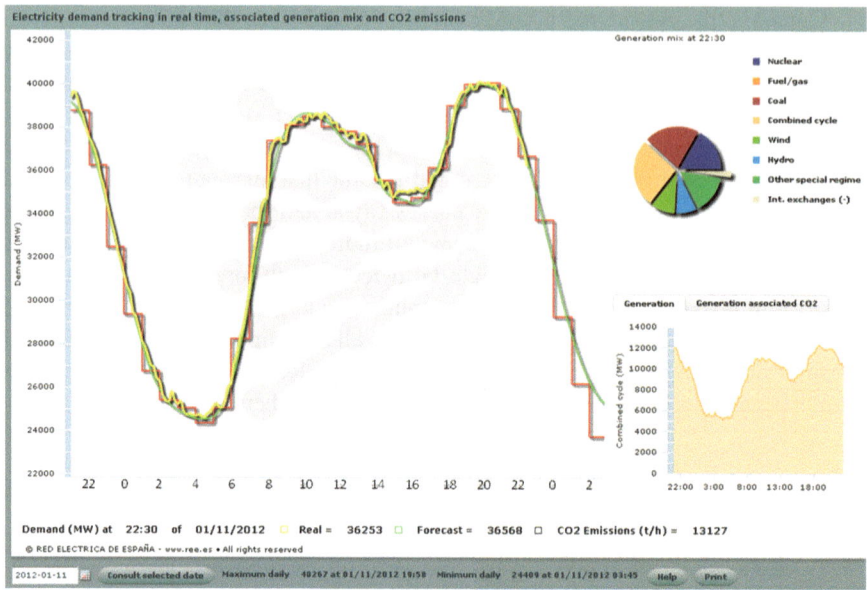

**Fig. 4.1** Demand/consumption and production curve and adjustments over one average day for Spain. *Source*: REE. For an online forecast, go to https://demanda.ree.es/demandaEng.html

systems, and today, it occupies a position of leadership in these activities. Their complete web page (also available in an English version) offers plenty of information on the status of the Spanish electric grid, including:

- The power demand and supply by type of energy for the Spanish electric grid, with online information for energy production and maximums (peaks) and minimums from each source of energy organized by year, day and even for every minute mix (Fig. 4.1).
- A list of the energy capacity installed and production for each type of power plants in Spain.
- An excellent updated 2009–2011 Annual Technical Report of REE, which provides complete information of electric power balance and installed capacity for all peninsular and insular systems in Spain.

# Literature

Hall, C.A.S., M. Lavine and J. Sloane. 1979. Efficiency of energy delivery systems: Part I. An economic and energy analysis. Environ. Mgment. 3 (6): 493–504.

Grandall, L., C.A.S., Hall, and M. Hook. 2011. Energy return on investment for Norwegian oil and gas in 1991–2008: Sustainability: Special Issue on EROI. Pages 2050–2070.

Murphy, D., C.A.S. Hall, C. Cleveland, and M. Oconner. 2011. Order from chaos: A Preliminary Protocol for Determining EROI for Fuels. Sustainability: Special Issue on EROI. 2011. Pages 1888–1907.

# Chapter 5
# Methods: Calculating the Energy Output.
# The Energy Returned (ER or $E_{out}$)

This chapter focuses on the methods used to calculate the actual energy delivered to the Spanish grid (the energy returned, ER or $E_{out}$) as used in the derivation of EROI. Our analysis begins with the fact that any energy conversion is associated with some energy *loss* as heat according to the second law of thermodynamics, so that the output has to be decreased by this heat loss factor, and that the generation of any equipment or service requires at least some energy. We use or occasionally sum these *losses* for each step in the process.

## Definitions of Solar Photovoltaic Power Used in Our EROI Analysis

*Nameplate power or* (*peak power*) (watt nameplate, Wp, or multiples such as kWp or MWp) of a PV plant is the power estimated by the solar PV modules' manufacturers. In order to calculate an EROI for a solar PV plant, we need to take into account the nameplate power, rather than the nominal power in the numerator.

*Nominal power* (watt nominal, Wn, or multiples such as kWn or MWn) for a Solar PV plant in Spain is the actual power output as legally measured by the government at the output of the DC/AC inverter. The highest payable tariff was originally awarded by the government to PV plants with an upper limit of 100 kW, as measured at the output of the inverter. This was done to encourage small entrepreneurs.

With these definitions and legal restrictions, virtually all the developers have installed modules that deliver much more power than the registered nominal power, typically ranging from 105 kW and up to 130 kW, the latter mainly in fixed (without trackers) plants. In this form, they can get 100 kW at the output of the inverter for many more hours in a year because the inverter has electronics inside to reduce the energy output to this level when necessary by changing the electricity to heat, which is vented. This occurs despite the already described losses from the theoretical peak power of the modules to the output of the inverter. This is done, in order to maximize

P.A. Prieto and C.A.S. Hall, *Spain's Photovoltaic Revolution: The Energy Return on Investment*, SpringerBriefs in Energy, DOI 10.1007/978-1-4419-9437-0_5, the number of legal limit generation hours in a year that will be paid at the best tariff rate, without exceeding the legally mandated maximum output of 100 kW per unit at each PV plant. We call this power at the module level "nameplate power" or "peak power," as specified by the manufacturers of the modules and the government's restrictions on the energy output that can be generated.

A Royal Decree/Law 14/2010 (urgent measures to correct the tariff deficit in the electric sector) recognizing this differential established an upper generation limit for the feed-in PV plants to adjust them into nominal values. This decree ignored the peak values of units installed, even if they were not producing more than 100 kW at the output of the inverter. In 2011, a new Royal Decree was enacted as a measure to reduce the government expenses in premium tariffs for the next 3 years. In exchange, they offered the suppliers to extend the time for which the premium tariff rates were granted from the previous 25 to 28 years. Businesses that had supplied the photovoltaic sector were angrily accusing the government of legislating backward and calculated the retroactive losses of expected revenues between 10% and 15%.

All solar production processes have a "nameplate" output. This output usually represents some kind of maximum output under what are often ideal laboratory conditions or at least in highly controlled environments. This is what is known as "peak power" of a given solar PV module. Actual output tends to be lower than the nameplate or peak power output. In our energy output analysis, we calculate what is known as the *Performance Ratio* (PR), the ratio of the losses that a solar PV system may have from the module to the invoicing point (usually the low-tension digital meter) compared to the electricity generated by the module (Fig. 5.1). The output factors we analyze represent the real (measured) photovoltaic system *losses* experienced in Spain under actual operating conditions and by the time the electricity is delivered to the grid, that is, the point of actual inventory (called in Spain the "border point"). This is what any sensible EPBT, LCA, or EROEI study should consider, rather than nameplate or peak power capacities with losses to the low-tension digital meter, as is often done.

We calculated the PR for a PV system in Spain from a number of known and tabulated empirical factors that are normally considered in every supply, installation, and commissioning contract in the PV turnkey plants in Spain. These factors vary for each plant, depending on technology, latitude, maintenance history, and through time. For Spain, a well-regulated and developed country, with good infrastructure, good engineers, skilled and experienced labor, and with adequate economic and financial resources, the actual output drops from 78% to 82% of the nameplate output. Even a stringent financial investor would consider 0.80 as a very acceptable PR. Most of the initial contracts were accepting bids with even lower performance ratios.

The *loss* factors that reduce nameplate power are listed as $l_1$ through $l_{12}$ in Fig. 5.1. In addition to these 12 factors, we have calculated other losses usually ignored in conventional analysis. These are specific losses before the electric energy is delivered to the grid throughout the total life cycle. These are listed as $l_{13}$ through $l_{15}$ in Fig. 5.1 and are developed more explicitly in the following section.

The main *losses* reducing the output for solar PV plants are shown in Fig. 5.1.

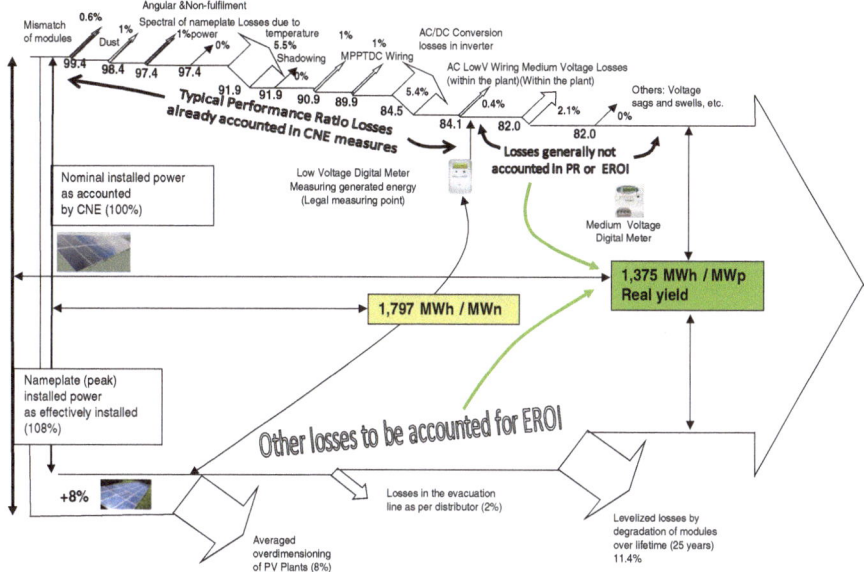

**Fig. 5.1** Sankey diagram of solar PV energy in Spain

## Annual Electrical Output Compared to Nominal Capacity for Spain

In 2009, Spain generated 6,074 GWh from its PV plants (corrected for performance ratio losses) from the average installed power base of 3,397 MWn; in 2010, Spain generated 6,405 GWh from the average installed power base of 3,619 MWn and in 2011, 7,398 GWh were generated with an averaged installed power of 4,043 MWn (Table 5.1).

## Factors Considered

### $l_i$: Mismatch of Modules

There is a regular conversion efficiency loss due to the mismatch of modules. The photocurrent of a string of modules connected in series (as they usually are) will be restricted by the least efficient module. The actual photocurrent will be lower than the value calculated from the input irradiance and the module's rated output current. This is especially an issue in installations of solar arrays where modules of different types have been assembled in the same array or when the same manufacturer has not carefully classified the modules for a specific array in series with exactly the same characteristics.

**Table 5.1** Installed feed-in solar PV power (in MW) and generated energy (in GWh) in Spain in 2009, 2010, and 2011

| As per CNE report May 2012 | | | |
|---|---|---|---|
| Year | Averaged installed power in MWn | Generated energy in GWh | Averaged generated energy in MWh/MWn per year |
| 2009 | 3,397 | 6,074 | 1,788 |
| 2010 | 3,619 | 6,405 | 1,770 |
| 2011 | 4,043 | 7,398 | 1,829 |
| Compound 2009–2011 | 11,059 | 19,873 | 1,797 |

*Source*: CNE report on May 2012 and own elaboration

Irregular shading, ice, or dust in modules of the same string also causes the mismatch in itself because it changes the characteristics of one module with respect to others in the same array. Differences in the short-circuit current and open-circuit voltage, as per the physics of the PV modules, or behavior of the bypass diodes may also create some mismatch.

Mismatch of cells is analyzed as one of the possible causes of the factor called "nonfulfillment of nominal power" of a given module when the sum of individual cells that compose the module is not perfectly matched or a cell suffers an unforeseen shadowing (e.g., a stork dropping). We use a factor for the industry of 0.6% loss, which is a commonly observed value ($l_1 = 0.994$).

## $l_2$: Losses Due to Dust

Losses due to dust are usually assumed to be 1% of all generated energy, but in field tests, especially in dusty regions (generally associated with well-irradiated regions in, e.g., Southern Spain), they are often as high as 4–6% if the washing and proper maintenance is not undertaken on a regular basis or the contracts do not include the washing of modules after a dusty rain or a dust storm or neighboring agricultural activities such as plowing. Some of the plants have been placed close to motorways with quite a lot of pollution or close to roads with gravel, which also add dust when transited. Even inside the plants, there are many paved roads adding dust when wind blows or machinery moves in for maintenance. High-concentration photovoltaics (HCPV) may have losses greater than 10% , due to the light dispersion in the front lenses, which is a serious burden for these technologies. Some promoters have observed that in some cases, dust may produce a positive effect, when temperatures are very high, by protecting the module from excessive temperature from irradiation. This can compensate for the losses caused by high temperatures generally important in sunny but naturally hot places.

Modules are rarely washed sufficiently or when needed, since contracts in the best case foresee two or four washings a year that are practiced either on fixed dates or when it fits into the schedule of the cleaning company. The collected data from 2009 to 2010 are not extensive enough to assess long-term figures accurately, so we assume a conservative value of 1% for this factor ($l_2 = 0.99$).

## $l_3$: *Angular and Spectral Losses*

This factor refers to the energy loss due to the reflection of the incoming irradiance at the PV module's surface when it is not pointed directly at the sun. This is calculated using geometrical optics theory. We use for this factor the conventional loss applied in most of the PR calculations in most of the projects for modules with antireflective coating and the effective refractive index, a conservative 1% ($l_3 = 0.99$).

## $l_4$: *Nonfulfillment of Nameplate (Peak) Power*

The nonfulfillment of the nominal power has disappointed many users, especially when buying in times of need. This occurs in the modules whose individual cells have not been carefully classified to exactly the same range of technical specs, in a production line that produces and stamps cells with a Gauss distribution of different efficiencies, or when the cells are not properly installed within the modules.

For example, we have had access to a published report of a specialized company that measures actual versus nameplate output (Coello et al. 2010, PP Enertis Solar). This company undertakes quality control of PV modules through visual inspection, peak power measurements, electrical isolation, and thermographic inspections. They selected modules randomly and analyzed 1% of all the modules from several providers for two big PV plants in Spain of 19 and 13 MWp. The actual modules were made by six different module manufacturers from different origins. The discouraging results indicate that:

> Peak power measurements show worrying results in both PV plants with manufacturers with extremely high percentages of modules with peak power below tolerance, 54.0% and 62.5% in Plant-A and 98% in Plant-B.

This situation has been improving gradually, after the demand collapsed in 2009 and because the learning curve forced many institutions to improve their measuring tools. This also convinced some big investors of the necessity for seriously inspecting the modules, once the hurry for reaching the deadline for the higher tariffs had passed.

Nowadays, it is common to find in some manufacturers new series of modules that guarantee −0/+3% tolerances about nominal power. We have assumed, again, a very optimistic position, with no losses attributed to lack of compliance with the nominal power ($l_4 = 1.00$).

## $l_5$: Losses Due to Temperature

Unfortunately for the solar PV systems, it normally happens that a higher irradiance is associated with high temperatures, especially in the longest days of the summer. Of course, there are some regions such as mountains or high plateaus that may have a high or very high irradiation and at the same time moderate temperatures, but this is not the case for most PV installations in Spain.

The losses due to temperature are sometimes surprising: in some supposedly good irradiance areas in Southern Spain, high or very high temperatures have proved to affect production significantly, specifically in the highest irradiation/generation months in summer. This is the case of regions in Spain, such as Murcia, Almeria, the plains in Castilla–La Mancha, and many places in Andalucía and Extremadura. Some initial specifications referred to peak power in such stringent conditions (1,000 W/m$^2$ of direct radiation and 20 °C temperatures) that could not possibly be achieved. This has caused some promoters to sue manufacturers and ask the government for arbitration, when they suspect that manufactures are not meeting the committed technical specs.

We use a very common value observed in most of the PR studies of 6.9% on the nominal output production at 25 °C or lower ($l_5 = 0.931$).

## $l_6$: Shadowing/Shading

Almost any random tour in Spain will find many PV plants that can be seen from the roads and motor ways that are installed in downhill positions that are not the most favorable. In some places, trees are shadowing some modules in certain hours of the day. For example, in an installation in Barquilla de Pinares, Cáceres, the plant was installed behind some eucalyptus trees. One year later, they had to cut the trees down due to the modules' lack of performance. In other areas, a mountain nearby may project a shadow on a valley where a PV installation has been located.

The greatest shadowing losses occur with mobile PV systems with trackers. We have analyzed Spain's 3.5 GW installed base and have found losses of 13% in PV installations with one-axis tracking system and as much as 24% in PV installations with two-axis tracking systems. In Spain, the lack of land in many cases and the interest in optimizing the energy capture of a legally limited plant of 100 kW have created more problems. This has led developers to build plants where the shadow projected by some trackers in some periods of the day (especially near sunrise/sunset) in the East/West orientations or even in some periods of the year (winter) in the South orientations generates shading losses as high as 2–4% of the daily output. This problem is aggravated because in many cases, manufacturers and EPC companies have not designed the installations carefully to minimize shadows. Even if the shadow of one tower with a tracking system projects only on a part of another tower, we may have losses higher than the theoretical masked surface. However, since most of the feed-in systems in

Spain are on the ground rather than on rooftops, where shadowing of mounted installations is much greater, and because most of the modules are overbuilt, we use a very conservative approach and assume that there are no shading effects ($l_6 = 1.00$).

## $l_7$: Maximum Power Point Tracker

There is a trade-off in energy more generally between rate and efficiency known as the maximum power principle that operates in electrical and also ecological systems (e.g., Howard Odum's maximum power principle in ecology). Maximum power point tracking (MPPT) is an electronic system that operates the photovoltaic (PV) modules in a manner that allows the modules to maximize the power (power, in Watts, equals voltage times amperage) they are capable of generating. The electronics in the power conditioning system (PCS) sometimes intentionally shift the operation point from the MPPT to a suboptimal operation point. This makes the voltage higher than the voltage that would deliver maximum power in order to reduce its output current, in order to prevent overvoltage to the power distribution line. PCS sometimes cannot find the MPPT point due to the difficulty in finding the right position on the current–voltage curve when it is very steep. When the capacity of the PCS is smaller than that of the PV's array, the output current will be restricted around the PCS's maximum output. Then the PCS will not track the MPPT and will keep the voltage constant if the irradiance is very low. We use a conservative 1% loss in the use of the device ($l_7 = 0.99$).

## $l_8$: Direct Current Wiring

Losses from wiring and mismatching increase if maintenance is not dedicated to fixing bolts and nuts from time to time. Spain suffered one of the first human victims of such green technologies when a PV module flew away during a wind storm, mortally wounding a person.

Thus, connectors when they are not well fixed or maintained by tightening them from time to time degrade electrical output. In general, direct current (DC) wiring is used between connections in the back of the solar panels and the inverters. The tables for copper cable diameters, as a function of the distance and power to be transported, are well known, and the protections and insulating devices are well tabulated, but developers do not always comply with the specs. This lack of compliance happened sometimes during the rush period of 2008–2009. We assume that this factor might increase as the projects age.

In big PV plants, the housing for the inverters used to be located as close as possible to the PV module arrays because losses in VDC are higher per meter and per diameter than in VAC. We assume a 1% in losses for this factor based on most common Performance Loss considered in commercial contracts ($l_9 = 0.99$).

## $l_9$: Alternative Current/Direct Current Output of Inverter

Inverters are fully electronic power devices able to transform 300–1,000 VDC output of solar cells into an AC voltage that in Spain is usually three phase 400 VAC. Today, inverters are astonishingly efficient with theoretical efficiencies reaching 97.5% of the input energy in DC. This is due mainly to the modern devices carrying the work of chopping a DC signal into square signals and taking the harmonics of these signals to the 50 Hz in an almost perfect VAC sinusoidal fashion. This is made possible using the "high-voltage insulated gate bipolar transistor" (IGBT). However, this has had a costly learning curve; as we have noticed in the early years, many IGBT's burned out during the summer within the containers mainly due to high temperatures and lack of sufficient ventilation. Losses in inverters are sometimes higher than nominal due to overheating depending on the air ventilation designs. In some cases, this may force the inverters to stop operating where the sun is in excellent position for irradiation. This has usually occurred in PV plants that are over dimensioned on the nominal power (i.e., 120 kWp of installed power in 100 kW nominal plants) or with deficient ventilation or cabins.

In some other cases, the module arrays were not properly designed for the range of operation of their inverters. As a consequence, they had to be reprogrammed to what is called "band's adjustment," thus decreasing the efficiency on the declared nominal power. We use a 5.4% loss for this factor based on standard values ($l_9 = 0.946$).

## $l_{10}$: AC Wiring Within the PV Plant

There are losses in the low-tension alternate current (AC) from the output of the inverter to the transformer's respective housings. This depends on the quality of the cables and connections utilized. We use a conservative loss of 0.4% ($l_{10} = 0.996$).

### Extended Performance Ratio Factors

While the performance ratio includes some important corrections, they occur before the plant output is officially monitored. It misses a number of other corrections that occur in the field after the output has been measured but before the electricity is joined to the grid. Specifically, every PV plant needs to have a low-tension AV sealed digital meter, which is controlled and monitored remotely by the electric operator (see Fig. 5.1). These meters are usually located next to the low-tension (400 VAC) side of the PV plant transformers. These meters keep a record of the energy produced in each quarter of an hour and store this information in their internal memory. This can be accessed remotely by both the owner of the plant and the electric power operator, either via Internet IP access or GSM/GPRS/UMTS data modems. In the case of solar PV parks or farms, where a number of individual plants of up to 100 kW are connected to the grid through a common line, it is also compulsory to install a second sealed digital meter that records all the energy delivered by the solar PV park or farm. This meter is usually located close to the connection and access point of the medium-tension line. In the case of solar PV farms, which are composed of a number of individual plants, there is another digital medium-voltage meter that sums all the energy generated by the individual plants after the transformers have taken the energy to some 15 or 20 kV (occasionally from 11 to 13.2 kV).

Hence, we derived what we call an "extended performance ratio," a concept that usually falls outside the conventional performance ratio (PR) losses used by the industry. We must not leave them out of any comprehensive net energy analysis such as an EROI, an EPBT, or an LCA. Below in factors $l_{11}$, $l_{12}$, $l_{13}$, $l_{14}$, and $l_{15}$ are the main reasons why we should include it although some promoters, investors, EPCs, or equipment suppliers are not especially focusing on them because they occur after the low-tension digital meter, which is the official source of economic revenues, although before the electricity is delivered to the grid.

## $l_{11}$: Medium-Voltage Losses (Within the PV plant)

### Losses in Conversion from Low to High Voltage

Utility companies usually subtract 3% from the low-voltage meters in order to calculate the energy to be invoiced. This is because there are additional losses in the step-up transformers that increase the voltage from 400 to 15 or 20,000 VAC for the high-tension lines (see Fig. 5.1). It is a common and legal practice that the electric power utilities automatically decrease the electricity delivered to the low-voltage meter (where the output is commonly measured) by some 3% to account for these losses. Losses between low and medium-voltage meters are in the range of 1%. This is not accounted for in the statistical data of the CNE, only in the final invoices. We have observed that small PV plants may have smaller losses, as the distances within the plant tend to be smaller. But if we include the transformer's losses, these PV plants will have a 2.1% loss ($l_{11} = 0.979$).

## $l_{12}$: Voltage Switch Offs, Voltage Sags, and Voltage Swells

The switch offs of the network (brownouts and blackouts) are not considered in the usual evaluations, but they are a real part of real supply systems. When switch offs or microswitch offs occur, solar PV equipment may be left idle while maintenance workers restart it.

PV plants in Spain are usually located at the end of unidirectional electric grids. These PV plants are not allowed to operate when the region in which they are located suffers a disconnection. The reason is to avoid the electrocution of servicemen when they are attempting to repair the fault in that zone. The disconnection is made by remote control, which is compulsory to have in each plant, by the utility company that has full control of the distribution stations (centros de seccionamiento u órganos de corte en red). In 2010, this was reinforced by new government regulations, for all PV plants greater than 2 MW.

In Spain, this does not happen often, but, for example, in Cáceres, a rural area in Western Spain, we have records of microswitch offs provoked in the power line by the local electric power utility. This affected a 1 MW PV plant with two-axis trackers preventing it from generating 6,000 kWh in 1 year alone. This is about a 0.31% loss of the generation of about 1,900,000 kWh.

Voltage sags and swells can represent a larger loss to PV plants. In May of 2010, in this same PV plant in Cáceres, a voltage surge of over 500 VAC in low-tension circuits was an immediate reaction to the maintenance team, something that is unusual in many plants with rectifiers for the protection of medium-tension systems. The damage amounted to 9,624 euros.

It is worth noting the lack of knowledge or interest that many LCA's show when trying to export these technologies worldwide to countries where brownouts and blackouts are common and frequent. This can have the result of seriously exaggerating the probable output of PV plants for these nations. Even in today's developed countries, crumbling societies in recession may lead to growing brownouts and blackouts. [However, no such theoretical losses are considered here ($l_{12} = 1.0$).]

## $l_{13}$: Peak Versus Nominal Installed Power Factoring

As we said at the beginning of this chapter, virtually all the promoters have installed modules that deliver much more power than the official registered nominal power (MWp), typically ranging from 105 up to 130 kW for the nominal 100 MW unit, the latter mainly in fixed (without trackers) plants. In this form, they get 100 kW at the output of the inverter (the inverter has electronics inside to reduce voltage to the legal limit of 100 MW). This allows the companies to get the maximum allowed 100 MW for far more hours in a year that will be paid at the best tariff rate, without exceeding the legally mandated maximum output of 100 kW per unit at each PV plant. We call this power at the module level "nameplate" power or "peak" power (kWp or MWp), as specified by the manufacturers of the modules.

The differences between the nameplate installed power and the nominal installed power are the result of having installed more MWp at the module level. These are on average 8% higher than MWn, which are used by the CNE in their tables. We consider this a very conservative figure, in a moment where the industry claims having losses of revenue of up to 30% for 90% of the installed PV plants in Spain due to the Royal Decree (Royal 2012). Certainly, there are two reasons for claiming these losses: one is the differential between peak and nominal installed power and the other is higher irradiance than the one assumed by the government when limiting the number of hours at peak per year. But the latter is far from being a 20% difference. Therefore, the 8% considered here as a differential between nominal and peak power is very conservative.

The excess of power at the input of the DC/AC inverter is beneficial for the PV plant, when the premium rates are high and sunlight is not at its maximum. But this has collateral consequences, since much more heat needs to be dissipated by the inverters, especially during hot summertime midday periods (and it is energy that is thrown away). This causes the inverters to either switch off automatically for excess of temperature, blow out, or for some key components to be burnt.

This factor is not included in some conventional LCA's, which take for granted the nameplate power as specified by the module producers/manufacturers. We included this factor in what we have called extended performance ratio factors. We assume a conservative 8% loss from the extra power installed in modules, a factor that is probably mostly applicable only in Spain ($l_{13} = 0.92$).

## $l_{14}$: Losses in the Evacuation Line to the Electric Network

Medium-voltage lines have their own electric losses from the compulsory terminal equipment installed in the PV plant (Centro de Seccionamiento u Órgano de Corte en Red—OCR) to the substation of the existing electrical network infrastructure. These losses are additional to the losses in medium voltage considered in factor $l_{11}$. They represent the losses beyond the digital meter in medium tension/voltage and the access or connection point.

Electric power lines have energy losses for several reasons. The most common losses are the joule effect and the corona effect. They are basically a function of the distance, the section, and the materials (copper, aluminum, etc.) of the transporting cables and insulations. These factors are influenced by the laying of the power lines (aerial or subterranean). This, in turn, depends on the type of evacuation line. In some Spanish regions, such as Extremadura, a well-irradiated and vast region, with almost one third of the territory protected with national parks or by European legislation [so-called ZEPA (Zonas de Especial Protección de Aves) or LIC (Lugares de Interés Comunitario) zones], some power lines had to be buried, as mandated by environmental impact assessments. We will discuss more details on monetary and energetic extra costs of areal lines in the denominator (input) factors of the EROI.

Typically, the average distance of a 1–2 MWn PV plant to the point of access and connection to the existing (or already built) high-voltage lines is about 800 m. The decision of the electric power utility on what type of lines to install is usually mandatory and depends on many factors. One important factor is the load of a given power line, where the plant is intending to connect. They do not admit feed-in connections of power plants that exceed 50% of the nominal carrying capacity of a given line. We assume an average loss of 1% for these lines, exclusively attributable to the medium-tension connection lines.

For the evacuation lines, there are companies specialized in saving money in these losses by a continuous monitoring of the differences between the low-tension meters and the medium-/high-tension meter, claiming that they can (on a commission for their service) accurately measure the real losses and dispute the rule of thumb used by the electric distribution company. They claim savings of 2.29, 1.79, and 1.63 percent, respectively (Vector Cuatro 2012).

In the case of solar PV farms of 3 MWn or more, the electric power utilities used to give the access and connection point to a substation. However, because there are few plants next to a substation, the distances usually range from 300 m to 15 km. Although there are no fixed policies in this respect, electric power utilities ask developers to install, for instance, 15–20/45 kV transformers in existing substations at the developers' expense. For solar PV plants that generate more than 10 MW, the electric power utility or distribution company may request the installation of a specific substation, which is subsequently networked with the grid. For the purposes of this EROI analysis, we assume that a reasonable factor for the connection line losses (including intermediate equipment such as substations) would be 2% ($l_{14} = 0.98$).

## *Would Rooftop PV Mountings Save These Losses?*

It is worth noting that many NGO's and environmentalists who favor, for environmental reasons, rooftop-mounted installations do not realize certain facts. There is a common argument to reject the above type of losses by arguing that in rooftop-mounted schemes there will be no need to have evacuation lines. Germany, which has almost 30% lower irradiance than Spain, has made most of its installations in this roof-mounted form. But this has further decreased the energy generated by lack of proper tilting or orientation (even on top of the German Bundestag). Many other institutional buildings, trying to show their willingness to promote and support renewable energies, make low-efficiency installations in their building walls, where the tilting of the modules is always inadequate.

A +/−1 m 3D map of a big urban environment, like the ones used for urban LMDS telecommunications, for a capital city in Spain or elsewhere in Europe, allows the shadows being projected from one building to the others to be calculated easily. In many cases, the installations are using all the available roof sites while supplying relatively little power. The price per square meter of the occupied land is usually a much more important consideration than the sun projections or the ecological sensitivities. The result is that most of the roofs do not fit well with the best efficiency.

One of the biggest problems for rooftop installations in Spain is the so-called horizontal property rights. In Spain, most of the inhabitants in urban cities live in apartment blocks, with little rooftop area per person/housing living underneath. In order to make a PV installation where the property of the roof is common for all neighbors in the block is sometimes complicated. This requires complex administrative and investment procedures, which not all the owners in the building can agree to since not all of them have the same purchasing capacity. This creates the problem of sharing the generated common energy among the different owner's electric meters.

# $l_{15}$: Degradation of Modules over Time

Solar PV manufacturers generally do not provide degradation rates for PV modules in their commercial or basic technical data sheets. However, we have discovered that when contractors are forced to provide them in their commercial contracts, they indicate a 1% average degradation per year up to a maximum of 20% in the modules' theoretical 25-year life cycle.

Some producers give a guarantee for the first 10 years of a maximum degradation of 10% and the rest up to year 25. Some plants suffer a higher module degradation rate in the first year, what is a sort of infant illness when first exposed to the sun that stabilizes later. We have seen 3% degradation during the first year and then a smaller degradation that varies from one brand of modules to another and even within the same brands. We assume a conservative 1% yearly degradation for years 1–20. The averaged factor over a period of 25 years gives ($l_{15}=0.886$).

This loss factor assumes that no big disruptions or collapses occur in the Spanish society during the 25 life cycle of the PV plants. But this may be quite unlikely given Spain's recent financial troubles. In effect, the ASIF report mentions that 1 year of financial troubles in Spain has led the PV industry from having 41,700 full-time and 26,300 part-time jobs in 2008 to a much lowered 11,300 full-time and 2,600 part-time jobs in 2009. If the job destruction in just 1 year of financial difficulties, in the heavily fossil-fueled Spanish society, is proportional to the destruction of solar PV companies, it is likely that many of the contractors that built these PV plants may not be there to honor their contracts including the equipment guarantees (usually from 2 to 5 years maximum and a few with 10-year material guarantee for the modules). It seems unlikely that they will honor their contracts and solve problems in relation with their theoretical commitment to provide an output power guarantee for the modules' life cycle of 25 years.

These contracts are difficult to honor because the companies serving them have simply disappeared, vanished or bankrupted, as discussed before. It may seem simple to react and find a substitute, but the newcomers, always willing to replace the bankrupted company, of course do not take over previous commitments, and this represents an extra cost, which at the end translates to extra energy costs that are not accounted for here. Because this phenomenon is recent, there are no records at this time.

For example, if a supplier of dual-axis trackers disappears, it may happen that the lack of maintenance of the trackers, which in many cases were first releases or experimental ones, may fail and be blocked, without spares or maintenance. This may affect the production of some plants in as much as 20% of the estimated production, while the failure of the deteriorated trackers persists. The substitution or repair of these trackers is not simple, as the concrete foundations are specific for each model or the machinery for the sun tracking is proprietary and difficult to emulate.

It seems sensible to wonder whether companies with shortened life cycles themselves, which may be shorter than a decade, will have authority or credibility to sign life cycles for 25 years for its products. Replacement of modules or inverters is always possible, but in order to analyze an accurate EROI, these extra energy costs have to be included in the energy invested required to build solar photovoltaic plants in Spain.

## Cumulative Impact

The total effect of all of these factors ($l_1$-$l_{10}$ before the meter, $l_{11}$-$l_{15}$ after) then would be

$$l_1 \times l_2 \times l_3 \times l_4 \times l_5 \times l_6 \times l_7 \times l_8 \times l_9 \times l_{10} \times l_{11} \times l_{12} \times l_{13} \times l_{14} \times l_{15}$$
$$= 0.994 \times 0.99 \times 0.99 \times 1.0 \times 0.931 \times 1.0 \times 0.99 \times 0.99 \times 0.946$$
$$\times 0.996 \times 0.979 \times 1.0 \times 0.92 \times 0.98 \times 0.886$$
$$= 0.655$$

When all these additional factors ($l_{11}$-$l_{15}$) are considered for all the solar PV production in Spain in 2009, 2010, and 2011, the output is reduced from 1,797 GWh/MWn to 1,375 GWh/MWp effectively installed in Spain. These are the actual measured averaged values, which are considerably less than the usual nominal and advertised values. These values show that even for a country as irradiated as Spain, with all the modern PV technologies available in a free market, where solar PV promoters have been able to choose the best places to build their plants, the actual operating conditions mean that only about two thirds % of the nameplate or peak power installed capacity is actually delivered to the Spanish society.

Some of these factors are appropriate only for the special circumstances of Spain in 2009. We would have to eliminate these factors for analyzing other situations, although each should be considered.

With these conservative factors and summary figures of the energy returned by a PV system, the key ER data to be considered for the all the PV plants in Spain in the years 2009, 2010, and 2011 already averaged in the three years period and calculated at present values for all the life cycle of the plants is 1,797 MWh/MWn, or gross energy output, and in a more net figure, 1,375 MWh/MWp. We now precede to calculating energy investments (EI) in the next chapter. The conclusion is that the final NET RESULT of the generated energy in Spain is 1,375 Gwh/GWp installed.

# Chapter 6
# Methods: Calculating the Energy Input. The Energy Invested (EI or Ein)

## Analyzing the Total Energy Costs of Photovoltaic Systems in Spain

Traditional life cycle analysis and energy payback time studies on solar PV systems often give estimates that appear quite favorable—that the energy invested in building solar collectors can be paid back by the device within 1 or 2 years (e.g., Fthenakis and Alsema 2006; Fthenakis et al. 2011). If the collectors last for 25 years (the usual assumption) this gives a rough energy return on investment of 12.5:1–25:1, rather favorable ratios. But these analyses usually focus on only the core processes of ingot/wafer/cell/module manufacturing, although sometimes including the energy costs of the associated direct equipment (inverters, trackers, if any, and metallic structures). In reality, the factors required to be able to deliver energy to society from a given energy production system go far beyond that and are related to each dollar spent on the project. The impetus for this study was the observations of the first author while being the Chief Engineer on several very large solar projects in Spain, when he began to wonder what kind of energy was being used to provide all the services and other components of the PV system. This led him to a series of beers with the second author, whom, he knew had experience with that issue, and, finally, this book.

We use a systems approach to give a far more comprehensive assessment of the energy costs of all components and activities of the system required to deliver PV energy in Spain in 2008–2010. Our interest in undertaking this analysis, as indicated in the preface, is the first author's experience with directing and supervising the construction of several of Spain's large PV projects, where he had to sign for every euro spent. Without all, or at least most, of these expenditures, the photovoltaic systems simply just would not work, and each is associated with an increased usage of fossil fuel somewhere in society. To undertake our analysis we identify a number of factors (hereinafter named $a_x$) as "energy additions" that should be included in the denominator of the comprehensive EROI equation in addition to the costs of the modules. We assume that when money is spent energy must be used to give substance to that energy, so a comprehensive energy analysis must start with a comprehensive economic analysis (e.g., Cleveland et al. 1984; Hall et al. 1986, Hall and Klitgaard 2011).

We employ five general methods in analyzing the energy contained/embodied or spent on the equipment, goods and services related to the solar PV plants in Spain.

P.A. Prieto and C.A.S. Hall, *Spain's Photovoltaic Revolution: The Energy Return on Investment*, SpringerBriefs in Energy, DOI 10.1007/978-1-4419-9437-0_6, © Pedro A. Prieto and Charles A.S. Hall 2013

The first is the direct energy used on-site, that is, to run a bulldozer or light lights. The second is the energy used off site to generate the materials used on-site, e.g., a solar module. The third is for a complex array of materials and services actually used in practice to meet the legal and operational requirements of the project. The latter require assumptions about the energy costs of goods and services, and we devote a section to this process and its uncertainties. The fourth is the energy represented by labor's paychecks, and the fifth is the energy associated with providing financial services. We are interested in obtaining a comprehensive analysis, but recognize that some of the inputs we undertake are controversial. We believe that the first three are relatively uncontroversial and the last two more, so, so we provide a number of important sensitivity analysis in the conclusion to allow the reader to choose what he or she wishes.

What follows is our estimate of each of the above five categories of energy inputs for Spain for the year 2009. We estimate each individual energy cost as a series of factors $(a_x)$ that are the estimated cost for that entity in 2009. These inputs are then summed and then divided into the output of the project (in chap. 7) to derive the EROI following the procedures given in Murphy et al. (2011). The two last factors $(a_{25}$ and $a_{26})$ are considered separately using a fairly comprehensive sensitivity analysis, as they may be partially included with some of the three main energy inputs considered, although it is sure that not all these energy expenses are embodied in them. These main factors are:

1. *The energy used on-site.* Some of these, represented in Table 6.18 at the end of the chapter, can be easily measured directly in energy *(factors $a_1$ through $a_6$).* Deriving the energy associated with construction or some other activity is not difficult, although it is not precision science. Where material and energy quantities are known (e.g., for 1 kg concrete about 1 Gj or 1.94 MWh, or that 1 kg of steel requires about 32 MJ = 115 kWh), it is straightforward. Deriving energy estimates for other entities is more complex, as discussed in Chap. 4.

2. *The energy used off-site to manufacture* the ingots/wafers/ cells/modules and much of the auxiliary or collateral equipment, such as inverters, trackers, and metallic infrastructures (if any). This is what most of the conventional EROEIs, LCAs and EPBTs deal with except for a few times when a small factor related to a restricted Balance of System (BOS) is included. This does not cover most of the factors we deal with here (factor $a_7$).

3. *The energy required to provide all other on-site or associated factors for which there are only monetary values* (represented in Table 6.18). This includes transportation of parts, washing of modules, and so on *(factors $a_8$ through $a_{24}$).* In some few cases transportation could be measured directly in energy terms.

   For sensitivity analyses, we consider the energy used for labor and for financial services. Since this is less frequently done, we include a more comprehensive description below:

4. *The energy used by all labor*
   Are the wages paid to labor part of consumption or part of production? This has resulted in controversy for energy analyses in the past. Most life cycle analyses make the assumption that not all activities and (energy) expenses incurred by the

**Table 6.1** Primary energy consumption attributable to the Spanish population, to active labor and to occupied labor

| Statistical data for 2009. INE | | Attributable primary energy per year | |
| --- | --- | --- | --- |
| Human group primary energy liason | Population | In MToes | In TWh |
| Total population | 46,063,511 | 142.07 | 1,705 |
| Occupied | 19,857,000 | 142.07 | 1,705 |
| Jobless | 3,207,000 | | |
| Active (+16) | 23,064,000 | | |
| Inactive | 15,292,000 | | |

Data for labor force from the INE

employee in the PV sector should be allocated to the energy input of solar PV systems because this is consumption, and most economists do not include it as part of production. Sometimes, they argue that the total energy expenses incurred by the full-time labor in the PV business would be incurred in any case, even if they were not working for that industry.

We wish here to do a comprehensive analysis of all energy required to construct a PV system, so we believe this factor should be provided, for if there were no wages provided and were energy not available to give meaning to wages (i.e., to provide the goods and services represented by wages), the work of labor most certainly would not be available. Since most of the energy used in Spain comes from fossil fuels, this energy cost is essentially from the fossil-fuelled sectors. However, we have not been able to get good data on the number or payment to workers without running a serious risk of double counting, for when services are purchased, labor is included in the costs. Thus, we include labor only partially to avoid such double counting.

We do wish however to point out how such studies could be done. First, we derive the total number of active workers in Spain (Table 6.1). For the employed people in 2008, the Spanish Instituto Oficial de Estadística (INE) offers the following figures for labor, considering the so-called active workers (all workers able to work in Spain) and occupied workers (the previous figure, discounting the jobless): the Spanish population consumes 142.070 MToes/year to feed and satisfy the needs of 46 million people, including workers, jobless, families, and relatives, etc. included in this equation. Since Spanish workers presumably spend their paychecks on more or less the whole suite of goods and services produced by the Spanish economy, in rough proportion to all of them, then it is appropriate to estimate the energy represented by their paychecks when we know only the monetary costs. The formula to relate the energy spent in labor by the Spanish photovoltaic industry is based, then, on the following assumptions: every employed (full-time) worker utilizes 142 MToes/20 million workers) = 7.1 Toes of energy per year. Since 1 Toe = 12 MWhn (or 41.87 gigajoules), we have 86 MWh (297 Gjoules) used per employed worker per year. Given that there were 41,700 full-time and within them 26,300 part-time jobs in the Spanish PV industry in 2008 (say 28,850 full-time equivalents), this translates to some 28,550 times 86 MWh = 2,455 million MWh (or 8,838 TJ). Given that the yearly averaged power over 25 year lifecycle for the modules produced in

2008 (2,758 MWn) is about 2,758 MWn*1,797 MWh/MWn = 4.95 Million MWh, then the energy weight of the manufacturing labour is about 50% of all generated energy in the lifetime. We cannot simply add this to our sum of other energy inputs as it may already be partially embodied in other energy input factors considered and also because the best data of the industry itself could be considerably inflated for marketing reasons to show it as a job creation sector. We just state that important amounts of labor may have not been included in the factors considered.

5. *The energy required to support business and financial factors* $a_{11, 12, 13, 15, 17}$ and $_{20}$
Energy is required to support the work of financial investments, including heating and cooling offices, operating computers, operating automobiles and airplanes in the normal order of business, and so on. As given in Chap. 4, we assume that these energy expenditures are about one third of national means (as derived from data in the USA) and thus use about 2.39 MJ for every euro spent in that sector.

## Calculating the Energy Inputs

### Direct Energy and Material Inputs

*Capital Equipment.* This is for the energy used directly on site to prepare the site for the PV modules, inverters etc., and to operate and maintain the PV plants.

#### $a_1$ Accesses, Foundations, and Perimeter Fencing

There are various "civil engineering" works required to prepare the site for the installation of modules and related PV equipment. This is for foundations, especially in the case of installations with trackers and for the access roads. We use the installed power up to end of 2009 from ASIF (2009) as the reference for these calculations.

External Accesses

The access roads built from the nearest paved road to the PV plant vary very much from plant to plant, and it is difficult to generalize, but we do. For the external accesses, we assume as typical a road of 5 m width and about 30 cm thickness of gravel road (zahorra) compacted with a road roller. We assume 2 tons/m³ of compacted gravel, which implies 1.5 m³ or 3 tons of gravel per linear meter of access road. For all of Spain, we assume a conservative average of 100 m of external accesses per solar farm for each of the approximately 3,000 solar farms or parks. This is about 300 km of access roads to plants, implying some 450,000 m³ or 900,000 tons of gravel. This volume requires 90,000 trucks of 10 tons each, traveling an average of 30 km (60 km round trip) from the deposits for a total of

5,400,000 km at 0.3 l of diesel per kilometer, for a total of 1,620,000 l of diesel. At 10.7 kWh/l, this is 17.3 GWh in fuel for trucks. We double this to get the consumption of road rollers, shovels, pickups, and cars for personnel and also the energy cost for grinding, mixing, and preparing the gravel (zahorra) with its corresponding machinery and the loading movements in the deposits to get a rough but probably conservative total of 25 equivalent GWh for constructing external accesses.

Internal Ways

Almost all of the fixed plants use modules installed in one, two, or three vertical rows. These need almost no internal infrastructure or operations other than some earth movements to flatten the space and prepare ducts for cables and some land compacting, perhaps with a minimum layer of compacted gravel in narrow corridors of between 2 and 3 m width. One-axis tracker plants, usually tracking azimuthally, that need some more separation between rows require a little bit more preparation and heavier machinery to fix the trackers. We assume that approximately half of the 1,400 MWp of combined fixed and one-axis trackers in PV plants have prepared infrastructures and the rest have left the fields with no access infrastructure. We assume an average of 1,200 m of corridors with 3 m. separation between rows of modules per each MWp installed, including all other service roads to inverters, transformers, and distribution station housings and to the control center. Fifteen centimeters of compacted gravel or specially treated compacted land is adequate to avoid plants growing in between. This represents an equivalent of about 0.45 $m^3$ or about 1 ton of gravel per linear meter of internal service ways or passages. There is a total of some 700 MW × 1,200 m/MW = 840 km of such internal service passages for the total installed PV power in Spain up to 2009. Therefore, the material for internal service roads and ways has required some 378,000 $m^3$ of gravel equivalent or 840,000 tons. This volume would require 84,000 trucks of 10 tons each traveling an average of 30 (60 km round trip) km from the deposits, for a total of 5,000,000 km at 0.3 l of diesel per km. 1,500,000 l of diesel at 10.7 kWh/l = 16 GWh in fuel for trucks. Again, adding this to the consumption of road rollers, shovels, pickups, and cars for personnel and also the energy cost in grinding, mixing, and preparing the gravel (zahorra) with its corresponding machinery and the loading movements in the deposits, we estimate a total of 22 equivalent GWh for internal accesses.

The two-axis trackers need a much better internal access to enable heavy trucks to access close to the posts and foundations to install and maintain the equipment, even in winter time, when the mud may prevent heavy trucks from entering for several weeks, even if an adequate pavement is prepared. The distances between trackers vary depending on the capacity of them; there are towers with two-axis trackers from 3 to 25 kWp in the largest installations, for the HCPV systems. At the end, there are no big differences, in occupied space, because the smaller towers may have smaller distances among them, but more towers have to be installed to get the same installed MWe. However, the larger the towers with the trackers, the better have to be the accesses to support heavy machinery involved in installation and maintenance or repair and access to the trucks with the elements.

For the 908 MWp installed two-axis power, we assume an average of 4 m width and 20 cm depth of compacted gravel and 300 m length of ways per MWp. This implies 240 m³ or 480 tons of gravel per MWp installed for a total of 218,000 m³ or 436,000 tons of gravel. This volume will require 43,600 trucks of 10 tons each, traveling an average of 30 km and back from the deposits for a total of 2,616,000 km at 0.3 l of diesel per kilometer or 785,000 l of diesel at 10.7 kWh/l=8 equivalent GWh in fuel for trucks. Adding this to the consumption of road rollers, shovels, pickups, and cars for personnel, and also the energy cost in grinding, mixing, and preparing the gravel (zahorra) with its corresponding machinery and the loading movements in the deposits, we derive a total of 12 equivalent GWh for internal accesses.

The total embodied energy here for accesses is then 59 GWh (Table 6.2). This is not a significant amount, and it is generally a one-shot investment in the life cycle, although some plants have had to repair the accesses in less than 5 years, due to the destruction caused either by the continuous passage of trucks, vans, and cars and/or the combined effect of rains and water flows.

Foundations

The foundations for the solar plants on the ground have changed considerably in recent years (Table 6.3). At the beginning, most of the installed plants were with trackers that demand some considerable foundations. Even for the fixed plants that require much less foundations, concrete foundations were common at the beginning. A typical row of two 2 m² and 2 m high modules may have 2 concrete cubes of 40×40×40 cm³ per each 4 m of row for a total of some 1 m³ of concrete for each of 16 concrete cubes.

We assume for fixed systems about 1,250 m of double module rows per MWp installed, each of which has 2 concrete cubes per each 4 m, and this is 625 cubes per MWp in fixed plants or 39 m³ or 70 tons of concrete for foundations per MWp of fixed plant. Therefore, the total use of concrete for the 2,462 MWp fixed plants requires about 172,340 tons of concrete. The embodied energy of concrete is about 1 Mj/kg and considering here 1kWh = 3.6 MJ; this will represent 0.277 kWh per KG of concrete, or 0.277 MWh/ton, which gives a total for the concrete used for all fixed plants in Spain up to 2010 of 172,340 tons×0.277 MWh/ton=47.7 GWh.

This volume requires 11,260 cement mixers of 15 tons each traveling an average of 30+30 km from the deposits for a total of 676,000 km. At 0.3 l of diesel per kilometer, 202,700 l of diesel were used, and at 10.7 kWh/l, this equals 2.1 equivalent GWh in fuel for trucks. Adding the consumption of vibrating machines, shovels, pickups, and cars for personnel, and also the energy cost in grinding, mixing and preparing the cement and concrete with its corresponding machinery and the loading movements in the deposits, we derive an estimated total of 4.6 GWh for transporting concrete for fixed plant foundations.

There are many differences among tracking systems. A typical two-axis tracker on a single post of 25 kW requires a foundation of about 20 tons of concrete for each 25 kW tower to fix the 14 m steel column, which is built 6.5 m into the ground. This represents some 600 tons of concrete per each MWp. Other trackers mount a galvanized

**Table 6.2** Embodied energy in external accesses, internal paths and ways of PV plants in Spain

| Embodied energy in | Width in m | Thickness in cm | Total gravel (m³/m) | Total gravel (tons/m) | Length in m | Total no of PV farms | Total m³ of gravel | Total tons of gravel | Total diesel in GWh | Total others in GWh | Grand total in GWh |
|---|---|---|---|---|---|---|---|---|---|---|---|
| *Type of infrastructure* | | | | | | | | | | | |
| External accesses | 5 | 30 | 1.5 | 3 | 100 | 3,000 | 450,000 | 900,000 | 17.3 | 12.7 | 25 |
| *Internal ways* | | | | | | | | | | | |
| Fixed plants | 3 | 15 | 0.45 | 0.9 | 1,200 | 700 | 378,000 | 840,000 | 16 | 6 | 22 |
| One-axis plants | | | | | | | | | | | |
| Two-axis plants | 4 | 20 | 0.8 | 1.6 | 300 | 908 | 218,000 | 436,000 | 8 | 4 | 12 |
| Total | | | | | | | | | | | 59 |

**Table 6.3** Distribution of PV plants in Spain by type of tracking system

| Type of PV plant of the 3,619 MWn installed in Spain (2010) | | |
|---|---|---|
| Type of plants | | MWp |
| Fixed | 63% | 2,462 |
| One-axis trackers | 13% | 508 |
| Two-axis trackers | 24% | 938 |
| Total | 100% | 3,908 |

steel post for a 15 kWp tower on a concrete dado of $3.5 \times 3.5 \times 1$ m, which represents some 12 m$^3$ or 21.6 tons of concrete per each 15 kW. This makes a total of some 1,425 tons per MWp. Another design of one axis trackers of 5 kWp per tracker uses concrete dados of $2 \times 2 \times 0.8$ m $= 3.2$ m$^3$ or 5.7 tons of concrete per each 5 kWp. This gives a total of some 1,000 tons of concrete per MWp installed. The huge differences between the foundations required for fixed and tracking systems are due to the fact that tracking systems offer much more resistance to wind than the fixed systems and the mobile parts of the former are more critical in front of windstorms. In fact, all these systems need a defense mechanism to place the whole tower in abatement position when the wind speed reaches a certain level, varying from 30 to 80 km/h. The tracking systems were preferred at the beginning because of the theoretical higher performance per MWp installed (30–35% more) and because much more concrete per MWp was not an economic burden when there were premium tariffs.

We assume that a value of about 1,000 tons of concrete per MWp installed is an acceptable but conservative estimate for one- and two-axis tracker installations. Therefore, we attribute some 1,400,000 tons of concrete to these types of plants. This volume required 93,333 cement mixers of 15 tons each, traveling an average of $30 + 30$ km from the deposits. This is a total of 5,600,000 km of distance at 0.3 l of diesel per km., resulting in a fuel consumption of 1,680,400 l of diesel. At 10.7 kWh/l, this equals 18 equivalent GWh in fuel for trucks, which we increase for the consumption of vibrating machines, shovels, pickups, and cars for personnel and also the energy cost for grinding, mixing, and preparing the cement and concrete with its corresponding machinery and the loading movements in the deposits. This gives us a total of 25 GWh for transporting concrete for tracking plant foundations.

The embodied energy of concrete, calculated at 1 MJ/kg, represents 0.277 MWh/ton, to give a total for the concrete used for all fixed and tracking plants in Spain up to 2010 of 1,400,000 tons $\times$ 1.94 MWh/ton $= 387$ equivalent GWh plus 25 for transportation is 412 GWh (Table 6.4).

Canals

Most of the plants have to dig trenches to lay power, communication, and control cables (about 1 m. depth) for protection from thieves. The length of these trenches is about 2 km per MW installed. They include inspection hatches,

**Table 6.4** Embodied energy in foundations of PV plants in Spain

| Embodied energy in | % of type of plants | MWp installed up to 2010 | Concrete for each MW_p in m³ | Concrete for each MW_p in tons | Total use of concrete in tons | Total use of concrete in GWh in GWh | Other energy exp in GWh | Total energy expenses in GWh |
|---|---|---|---|---|---|---|---|---|
| *Type of infrastructure foundations* | | | | | | | | |
| Fixed plants | 63% | 2,462 | 39 | 70 | 172,340 | 47.7 | 4.6 | 52.3 |
| One-axis plants | 13% | 508 | | 1,000 | 1,400,000 | 412 | 25 | 437 |
| Two-axis plants | 24% | 938 | | | | | | |
| Total | | 3,908 | | | | | | 489.3 |

distribution boxes, or manholes that have to be watertight. There are also canalizations and earth movements associated with roads to control water flows, something quite common in most of the parks or farms that in some cases have had to proceed after being in operation to carry out corrective actions, when parts of the plants and equipment were flooded by rains. We assume, for the sake of simplicity, that these energy expenses are included in the foundations calculated above (Table 6.4).

Fencing

There are many types of fencing associated with PV stations, but according to the existing environment regulations in the most important communities, it is compulsory to have the so-called hunting closures, which in fact are good for small animals (quails, rabbits, etc.) to trespass but are bad for security, as they are much easier to break through than other metallic fences more appropriate for security, but forbidden in rural areas.

We estimated a mix of PV plants as in Table 6.5.

For estimating the energy costs, we multiply the lengths in Table 6.5 by the lightest type of fence, 2 m high and about 100 kg per 100 m roll (1 kg of galvanized fence per meter). This gives about 3,350 tons of galvanized steel utilized in fences and another 3,350 tons of galvanized steel in posts and accessories. Considering that 1 kg of steel requires about 32 MJ = 115 kWh, this would be about 385 GWh. This will also require about 670 trucks of 10 tons capacity to take the materials to the solar fields, traveling an average of $50 + 50$ km from the manufacturing sites to the plants. Foundations for the posts take approximately 1 m$^3$ of concrete for each 20 posts. This represents about 50,000 tons of concrete. At 1.94 MWh per ton of concrete this would be 97 GWh. This will require 5,000 trucks of 10 tons each, traveling an average of $30 + 30$ km from the cement factories to the plants. Without considering the trucks moving for the transportation of fences or concrete for fences to the sites, we have an approximate total embodied energy for all aspects for the fencing of $385 + 385 + 97 = 867$ GWh.

In conclusion, all the energy embodied in the accesses and in both the manufacturing of all concrete used for foundations and in the transport and installing is as follows:

| | | |
|---|---|---|
| • | Accesses and ways | 59 GWh |
| • | Foundations for the plants | 489 GWh |
| • | Fences, posts and foundations | 867 GWh |
| • | Total embodied energy | 1,415 GWh |

Or, if distributed throughout the 25 years, it represents ($a_1 = 56.6$ GWh per year) spent up front. This (and all subsequent such calculations) can be compared to the total electricity generated for all Spain for years 2009, 2010, and 2011, which averages 6,624 GWh/year (Table 5.1) gross energy output and 5,069 net energy output. Thus, the total yearly spent energy for $a_1$ is 1.1% of the energy output.

**Table 6.5** Approximate calculation of the perimeter fences of functional (not individual) solar PV plants or parks installed in Spain up to 2010

| | | Approximate plant size | | | |
|---|---|---|---|---|---|
| | | <2 MW 36% | 2–5 MW 20% | >5 MW 44% | |
| Embodied energy in | MWp installed up to 2010 | No. of plants | No. of plants | No. of plants | |
| *Type of infrastructure* | | | | | |
| Fencing | | 0.5 MW avg | 2.5 MW avg | 5 MW avg | |
| Fixed plants | 63% | 2,462 | 1,773 | 197 | 217 |
| One-axis plants | 13% | 508 | 366 | 41 | 45 |
| Two-axis plants | 24% | 938 | 675 | 75 | 83 |
| Total | | 3,908 | 2,814 | 313 | 344 |
| *Approx. perimeter fences per plant in m* | | | | | |
| Fixed plants | | | 500 | 600 | 800 |
| One-axis plants | | | 1,200 | 1,600 | 2,000 |
| Two-axis plants | | | 1,800 | 2,200 | 2,400 |
| *Total perimeter fences in m* | | | | | |
| Fixed plants | | | 886,320 | 118,176 | 173,325 |
| One-axis plants | | | 438,912 | 65,024 | 89,408 |
| Two-axis plants | | | 1,215,648 | 165,088 | 198,106 |
| Total perimeter fences: 3,350 km | | | 2,540,880 | 348,288 | 460,838 |

*Sources*: ASIF Report 2009 (page 12) for type and size of plants and own extrapolations for occupied land per plant and security perimeter per plant

**a₂ Energy Investments for Evacuation Lines and Their Rights of Way**

As the electric power lines and substations which the electric power utilities assign for the connection to the PV plants are seldom next to the locations where the PV plants are installed, there is a need to set up evacuation lines in between.

Spain has a regulation with respect to medium tension power lines (15–45 kV) that if the lines are considered of public interest, they may force expropriation of the land required for the foundations for the poles or columns and the fly over rights of the cables from the owners of the lands affected—with compensation of about 100 € per small steel pylon or pole of up to 20 m height. That process may take 1 year of heavy administrative procedures, even in the middle of the race to arrive on time for the highest possible subsidies. Then the developers try to reach private agreements in most of the cases with neighbors. In the beginning, the rural landlords were unaware of the PV business and offered their lands at cheap prices per column or pole. Sometimes, they bargained for a right to connect their own consumption point to the line, which in some rural areas is an advantage. However, soon the farmers realized that the business was significant enough to increase their fees. We know of some medium tension poles, which were key, that have cost up to 24,000 €. In some cases, evacuation lines have to be subterranean, for environmental reasons (Zones ZEPA or LIC for special protection of the protected fauna, such as storks, or predators, like eagles, or scavengers like vultures) or other reasons (railways passes, urban areas, thieves, etc.) In those cases, prices are about triple that for an aerial power line of the same capacity, depending on the land. Permits and rights of way are more complex because of the impact needed to excavate the trench along all the line.

We make the following assumptions to calculate these costs. From about 3,000 solar PV farms or parks in all Spain, we assume a conservative 2,000 require rights of way with an average of 800 m of power lines to the access and connection point (about 8 columns or poles with 100 m distance between poles) at an average price of 800 € per pole for a total of about 13 million euros. There is an additional economic and energy expenditure for the line materials themselves, usually of the order of 20 or 15 kV, which we assume cost about 35,000 €/km of evacuation line. Assuming conservatively that these materials require the same energy per dollar to produce 1.980 MW, the total economic costs for access lines amount to some 83 million euros, about 3.3 million euros/year, or as energy 0.1% of the equivalent energy output in the form of electricity.

## *Other Necessary Energy Investments Derived from Economic Expenses and Translated into Energy Equivalences*

### a₃ Operation and Maintenance Energy Costs

All components of PV systems, like just about everything else, require additional money and energy use for operation and maintenance (O&M). The contracts that are signed for maintenance purposes vary very much from some plants to others. When the suppliers are taking care of the whole or the most important part of the value chain (from the module, inverters and/or tracker's manufacturing, R&D, engineering, operations, etc.), the buyers of the plants prefer to have a long-term O&M contract that gives them more comfort.

Supply, installation and commissioning, or turnkey, contracts usually offer a guarantee for materials between 2 and 3 years to a maximum of 5 years for modules and inverters. Some modules were offering up to 10 years guarantees for materials for modules in the first deployments. When the demand was largely exceeding the supply during the period 2007/2008, they reduced the guarantee immediately to 5 years in most of the contracts. From 2009 onwards, several companies with large equipment stocks due to the financial crisis started again to offer 10-year materials guarantees for modules.

It is worth understanding the difference between the guarantee for the materials and that for the power output that are normally offered for 25 years. Many original equipment manufacturers (OEM's) for PV modules (or their representatives, resellers or distributors), when consulted, on the possibility of having a free of charge replacement of a faulty module if the failure falls 2 or 3 years beyond the 5-year guarantee for the materials, answered that the purchaser should acquire a new module and pay for it. The power output guarantee is subsidiary to the materials guarantee, and the latter supersedes the former, which is activated only if the PV module works normally but the power output falls below the specified output over time.

In this respect, even the modules, inverters, trackers, and auxiliary equipment have to be replaced free of charge within the guaranteed period, the O&M contracts usually start from the provisional acceptance of the PV plant. Many O&M contracts are signed for an amount that is a proportion of the total generated yearly income. In these cases, no updated or escalating formula is generally applied, as the premium tariffs are supposed to be regularly (yearly) updated by the Consumer Price Index (CPI). Some other O&M contracts fix a global yearly fee that is generally and automatically updated as it by the yearly updates of the CPI.

In both cases, a rule of thumb is that O&M contracts take between 5% (fixed) and 10% (two axis) of the total revenues. The total sales of energy from the PV plants in 2009, 2010, and 2011 was 8,527 million euros. We use a weighted average of 7% of the revenues for O&M, as the total yearly maintenance cost in monetary terms, which amounts to an average of some 596 million euros. This is 199 million euros per year. Using the national mean Euro to energy conversion ratio for 2009 (1,980 MWh/million euros), we derive $a_3 = 394$ GWh/year, about 7.7% of the energy output. Most of the maintenance contracts do not foresee replacement of materials beyond the material's guarantee period.

## $a_4$ Module Washing and/or Cleaning

Most of Spain is a dry, dusty place. There is not enough fresh water for all the needs, and considerable energy is required to pump, purify, or otherwise get water and to pump it to its different destinations, in many cases by means of transfers among different basins. Then more energy is used for treatment, sprinkling or for drop-by-drop cultivations. Only 4.16% of national water use is supplied by gravity. Spain treats water for some 54 million inhabitants, and this volume of treated water amounted to some 8.6 Hm$^3$/day or some 3,140 Hm$^3$/year (Miliarium 2008).

There is not enough natural fresh water so that some of the water must be obtained from desalinization of sea or polluted water. Spain is the fourth country in the world in desalination capacity, after Saudi Arabia, the United States, and the United Arab Emirates, producing on the order of 2 Hm$^3$/day (about 720 Hm$^3$ = 720 million m$^3$ per year), depending on the hydrologic year, with a total installed capacity in 2009 of 990 Hm$^3$/year. The energy cost for desalination has improved considerably, but it still requires some 3–4 kWh/m$^3$. Therefore, the total cost of desalination per year in Spain is about 720 million m$^3$ × 3.5 kWh/m$^3$ = 2,520 × 10$^6$ kWh/year = 2,520 GWh/year. We assume that all operations apart from desalination that make water available to society also require energy use of 3.5 kWh/m$^3$.

Considerable water is needed for cleaning modules. Even though it does not go through microbial purifying systems, it has to have some other expensive processes (sometimes more than five) to be acceptable for cleaning PV modules. Therefore, we assume a use of 7 kWh/m$^3$.

In Spain, modules are washed to clean them and remove the dust layer and other particles adhering to the surface with time. There has been a big learning curve. Some two-axis tracking installations initially used about 25,000 l annually of de-mineralized, decalcified, and, in some cases, deionized water per MW. At present, a 1 MW PV installation can be cleaned with about 5,000 l (5 m$^3$/MWp) of this type of clean water. In the years of massive deployments, there was no experience to know or learn about the advantages of washing the modules or the importance of periodic washing. The owners soon learned that cleaning was worth the cost in exchange for the losses avoided. Many plants were initially washed with water from the tap or wells or nearby ducts, with sometimes consequences of obscuring the panels, because calcium or heavily mineralized (salted) contents caked or even etched the glass. Inappropriate brushes or inadequate pressure of the cleaning systems scratched the glass. We shall not consider here the damages that occurred or will occur in the medium or longer term, as it is difficult to assess the problems happened in the origins of the learning curve.

The O&M contracts of large plants usually require a minimum of two washes per year to avoid losses. In many cases, this is completely inadequate, especially in the arid southern part of Spain (where the higher irradiance is), which has many dust storms and rains carrying plenty of mud. Sometimes, as per Murphy's Law, it happens that immediately after a washing of the park, a rain with mud leaves the park in a situation even worse than before the washing. Or the O&M contract established some fixed dates that did not coincide with the dustiest periods of the year or the plowing of a neighboring field. Sometimes it happened that specialized cleaning companies that surged after the boom and are localized for reasons of proximity and better service could not attend simultaneously to all the contracted parks in a given region, immediately after a rain with mud or a dusty storm has passed through a whole region where several plants are installed.

We assume here that there was an average of four washes per year, using 20 m³ of pure water per MW per year. This gives an estimated total of $20 \times 3,908$ MWp $= 78,160$ m³ of purified water per year for all the installed PV plants in 2010 in Spain. This is not much energy but represents 7 kWh/m³ of purified water times 78,160 m³ $= 0.5$ GWh/year. The cleaning teams going to the sites are usually composed of a truck or van with the cleaning equipment (sometimes, especially for one- or two-axis trackers, with a lift or crane or telescopic crane/arms) and the associated water tanker, usually 5–10 m³ capacity that makes, as many round trips as necessary, from the source of clean water to the site/s. We assume for all of Spain 8,000 round trips of 10 ton trucks for $30+30$ km average distance for 80% of the plants involved. That makes a total of about 480,000 km for trucks of 10 tons, using about 1 million liters of diesel on average. At the ratio of 10.7 kWh/l of diesel this is 10.7 GWh/year. The total energy consumption is about 11.2 GWh/year. This is 0.22% of energy output ($a_4 = 11.2$ GWh/year).

## $a_5$ Self-Consumption of Energy in Plants

The internal consumption of a plant is something that also needs to be considered. Fixed plants consume much less energy than plants with trackers, as the latter have to move the modules to be as perpendicular to the sun as possible. The energy consumed in a PV plant is generally composed of the following items:

- Self-consumption of trackers (some of them are electric motors used directly; some others, electric motors pressurizing a hydraulic system whose electric valves control the hydraulic arms—or different hybrid versions). Some of the consumption is for the electric fans evacuating hot air from the inverter's or transformer's housing. There are known offers for apparatus to ventilate/refrigerate units with electric fans to gain some net energy by lowering the temperature of them in very hot days (production can fall from 0.4% to almost 1% per degree Celsius above 25°C), or the inverter's housings. These formulas are not well developed.
- Control room. Most of the plants have a control room, manned, or not. Lighting, water pumping, sewage treatment, purifier, aircon, heating for winter, etc., are all requirements.
- Communications. They might be installed or not in the control room and are devoted to send generation data, climatic and other parameters, and technical alarms to the management centers, usually remotized in most of the modern big plants. In many cases, especially in summer, these control rooms have air-conditioning even when they are unmanned because the computer equipment and electronics inside may fail if they get too hot.
- Security systems. Some installations have proximity and/or movement detectors or laser or infrared beams to protect the installations and remote

**Table 6.6** Self-energy consumption by type of PV plant

| Type of plant | % | Installed power in MWn (2009–2011 avgd) | (2009–2011 avgd.) self consumption in kWh/MWn/year | Total national (2009–2011 avgd) self consumption in MWh/year |
|---|---|---|---|---|
| Fixed plants | 63% | 2,322 | 5 | 11,612 |
| One-axis trackers | 13% | 479 | 7 | 3,355 |
| Two-axis trackers | 24% | 885 | 15 | 13,271 |
| Total | 100% | 3,686 | 7.6 | 28,237 |

*Source*: ASIF Report July 2009, CNE data 2009 to 2011 and own elaboration

connection with security companies, in an attempt to avoid stealing of the equipment or vandalism. Guards may attend when the alarm sounds or they may be contracted with random patrols. Some of them have permanent watchmen. Some big installations have a perimeter security lighting switched on during nights or, more usually, in case of alarm.

• For the self-consumption of the functional elements considered (those necessary for the direct functioning of the plant, like motors for the trackers, electric fans in the inverter's housings and alike), the Spanish Administration has regulated that this type of consumption has to be subtracted from the electricity generated with premium tariffs, something which seems to be logical. For this purpose, digital meters allow bidirectional measurements of the net electricity flows.

However, for the rest of the electricity consumption of a PV plant, the electricity can be taken and consumed at market prices. This is taken from a transformer installed in the plant generally named "auxiliary services," with a specific digital meter.

We will not count the electricity that is self-consumed, but subtracted by the bidirectional meters, because it is already subtracted from the output reported by CNE and considered implicitly in our analyses. It is quite complex to know how much energy is consumed for the rest of functions that can be paid at market prices here, but that are necessary to operate the plant because it varies very much from some plants to others and by the type of plants (Table 6.6).

The consumption observed in two-axis trackers plants with permanent presence and frequent use of the air-conditioning/heating equipment in the control room, pumps, etc. in the auxiliary services digital meter, is typically of about 15,000 kWh/year/MWn. For one axis plants, we have observed some of them working with PV cells additional to the generating modules, charging a battery to power a small electric linear motors, push-pulling an arm. In this case, no electricity consumption has to be considered for the self-consumption of the plant, although this implies more auxiliary equipment, that is more cells and batteries; more energy that will not be considered here. We assume some 7,000 kWh/year auxillary power/MWn. For fixed plants, this consumption may be reduced to some 5,000 kWh/year/MWn.

Therefore, as per the mix of installed PV plants installed in Spain by 2011.

As we have seen, the net energy returned in 2009 for the total PV installed plants in Spain was 5,069 GWh/year averaging years 2009, 2010, and 2011. Then, this factor represents 0.5% of the energy output ($a_5 = 28.2$ GWh/year).

## $a_6$ Security and Surveillance

The cost of a contract with a security company may vary a lot from some PV fields to others. In general terms, small plants have unmanned security systems with movement detectors, or infrared or proximity or tamper devices, and they send the alarm signals to a security and surveillance center. A growing number of them have some cameras, with or without tracking systems and night vision in the perimeter of the PV field and/or in sensitive places (entrance, accesses, control rooms, etc.). They are sometimes complemented with high-capacity recording systems, sirens and perimeter lighting, the latter usually connected only in the case of alarm. The communications and remote control are explained in another section.

A typical security materials cost for a 1 MWn plant may range from 20,000 € to more than 100,000 €. Of course, the larger the plant, the lower the cost of the alarm systems per MW. Although nominally there are more than 50,000 PV plants registered in Spain, they are usually grouped in the so-called solar farms or parks in 100 kWn or smaller units assembled within the same perimeter and separated only by the cabling and meters, as per the legal requirements, to get the higher tariff. The distribution of PV plants by MW power generated is given in Table 6.7.

Depending on the contract, they have different obligations to attend to. Some contracts receive only the alarm signal and call back to the designated person of the owner to attend to the alarm. Some others, more expensive, have periodic, random patrols around the field a number of times per week, usually during night times. For the three types of combined plants we assume an average cost of materials of 10,000 € for the plants <2 MW, 25,000 € for the 2–5 MW combined plants and 50,000 € for the combined plants or PV fields >5 MW. These amount only for information purposes, as we are considering here basically turnkey projects that usually and contractually include these systems.

As for the specific security staff and means (not considered in the factor $a_{25}$ of direct + indirect Spanish labor) as sensitivity analysis nor in the factor $a_{16}$, of circumstantial and indirect labor, we assume that the big plants of >5 MW have a contracted security staff with permanent presence and patrolling within the plant at a cost of about 10,000 €/month (120,000 €/year) on 24 h and with personnel attendance on weekends . For the PV plants in the range 2–5 MW, we assume that half of them have personal presence at 10,000 €/month (120,000 €/year) and half at 3,500 €/month (42,000 €/year) for external remote surveillance with attendance in case of alarm. For plants <2 MW, we assume all have a remote maintenance which includes only a cheap and simple remote control and surveillance of 500 €/month (6,000 €/year). Therefore, we believe that a reasonable estimate of all total surveillance and security costs would be in the range of 70 million euros/year for all the plants in Spain. Using the national mean euro to energy

**Table 6.7** Distribution of combined PV fields by plant size in Spain

| Plant size average 2010 | | Installed power per size type in MWp | Estimated number of solar PV combined installations |
|---|---|---|---|
| < 2 MW | 36% | 1,407 | 2,814 combined plants of 0.5 MWn average |
| 2–5 MW | 20% | 782 | 312 combined plants of 2.5 MWn average |
| >5 MW | 44% | 1,720 | 344 combined plants of 5 MWn average |
| | | 3,908 | |

*Source*: ASIF Report 2009. Page 12 for sizes split in percentage, CNE Report of May 2012 for the 2010 individual installations and own elaboration for the rest

conversion ratio for 2009 (1.980 kWh per euro), this is equivalent to 2.4% of the averaged energy output from all the solar PV plants ($a_6 = 138.6$ GWh per year).

## *Indirect Energy Inputs and Material Inputs*

In addition to these relatively small inputs that we assign to on-site costs, there are many energy costs that occur off-site, most obviously to manufacture the PV modules.

### Energy Derived from Conventional Life Cycle Analysis Studies and Calculated as an Inverse Factor of an EPBT

$a_7$ Modules, Inverters, Trackers, and Metallic Infrastructure (Labor Excluded)

Although certainly there has been some progress in the improvement of the photovoltaic technologies and in some cost reduction programs associated with this technology, there is sometimes a mythology about the never ending cost reductions and efficiency improvements. We have observed through many real life offers, projects, and contracts price breakdowns that PV modules represent slightly more than 50% of the total project cost for a fixed plant for an investor. Many other LCA's or EPBT's give similar percentages, both if they use prices or energy embodied in the processes. One-and two-axis PV systems usually have a lower percentage content of PV modules in the total project for obvious reasons. We assume that modules and inverters are 60% of the energy content of the framed module in a complete PV system.

Most of the LCA's and EPBT's studies and papers dealing with the materials composing the PV modules show different percentages and treatments of the silicon feedstock or ingots + wafer + cells with respect to the framed module. Energy costs range from 3,500/4,600 MJ/m$^2$ module area in poly-crystalline (76% of all energy inputs considered basic to the solar PV modules or systems) to 800/1,600 MJ/m$^2$

(50%) in amorphous silicon thin film modules (Fthenakis, and Alsema 2006) or 800/1,200 MJ/m$^2$ (67%) for CdTe PV modules (Fthenakis and Kim 2011. The dispersion is higher, and the lack of standards to measure the energy contents in an evolving technology makes it difficult to produce an educated guess. Others, calculate the energy pay-back time, by breaking down into "Laminate, Frame and Balance of System" for different technologies (Fthenakis et al. 2011). Some others give price indications of the module contents and again split at convenience in "wafer, cell and module assembly," with results that range from 25% to 60% in wafer + cells content with respect to the total components costs of poly-crystalline silicon modules (A.D. Little 2001).

As most of the Spanish PV-installed base has been made on poly-mono and or single-multi crystalline silicon systems, with a heavier economic/ energetic content of the cells within the module, we assume that cells and their precursors (feed stocks, ingots, wafers) have an average of a 70% of the energy content of the module. Thus cells represent approximately 60% × 70% = 42% of the total embodied energy of the system (modules, inverters, trackers, and metallic infrastructure—labor excluded—as per the usual, and using the limited and restricted conventions of the industry analysis).

When a system claims a much lower cost (economic and/or energetic) than mono-poly or silicon modules, it is usually because it has a lower efficiency, as is the case of thin film, amorphous or CdTe modules, which obtain real efficiencies of one digit but often do so cheaply. The much higher efficiencies, such as the nominal 20% or even 40% of the multi-junction ones, are because their costs (both economic and energetic) have skyrocketed much higher than their efficiencies, forcing the users to invent concentrating systems or limiting their use to specific military or space applications, far from massive deployments, which is our main interest here.

We have even seen recent claims that new "photon enhanced thermionic emission" PV systems could reach up to 55% and even 60% conversion efficiency, or incredible promises about graphene. A new PV imaginary system that will appear in the market doubling the efficiency of present modules while theoretically keeping similar energy inputs or needs could in principle double the EROI. PV modules bought by installers have gone from 12 to 16–17% photon to electricity conversion at the cell level. At the module level, we have to discount some 1% or 2% for mono-poly and/or single-multi-silicon modules.

For the sake of simplicity we shall not spend too much time on this factor and shall use the EPBT's of the industry in its more conservative (optimistic) side, an EPBT of 3 years for this $a_7$ factor, as applied to the Spanish situation as of about 2008. Life cycle is universally considered 25 years in all these studies (there are few adventuring some 30 years), so this means that the modules and the restricted Balance of System as per the conventional EPBT studies (labor excluded) recover 25/3 = 8.33 times more energy than what they consumed in its life cycle. Given that in 2009–2011 the average net electricity produced was 5,069 GWh, the energy invested up front, but averaged over each year, for this factor, was 1/8.33 (or 12%) of the energy output or 608 GWh/year ($a_7 = 608$ GWh).

$a_8$ Transportation: from Local Manufacturing to Air Shipments from China

Some existing LCA's (e.g., Fthenakis and Kim 2006) mention the transport issue in a Balance of System (BOS) of a photovoltaic system. Their numbers are always very small (e.g., 10 Mjoules of energy per square meter of PV module in a total of 542 $MJ/m^2$ and some 331 $MJ/m^2$ for frame total (Fthenakis et al. 2011). This is only 1% of the module or PV system.

We offer some more detailed information on the economic and energy expenses of the transportation from our experience in Spain. We start by analyzing the processes in Spain and compare it to this global world. And then we take a look to the main processes of this PV sector.

Spain imported in 2008 far more modules than it produced (Table 6.8) generally from very distant places, mainly for economic reasons. The data about the production of ingots and cells (Table 6.9) indicates that most of them were imported from the world microchip technology sector (mostly in Japan, the USA, and Germany but increasingly China). Spain invested in only a few areas and at exactly the worst moment, when the domestic market collapsed. Table 6.9 gives our very conservative assumptions on the origins of the different parts of the solar PV business to supply the booming 3,668 MWp installed solar PV power in the Spanish market in 2009. Most of the PV systems must travel several thousand km to reach their destiny.

Ingots, for example, are produced by a very few companies located mainly in Japan, the USA, and Germany (Table 6.9). Most of ingot/poly-silicon manufacturers are so specialized that they are not in other segments of the business, so the ingots must be shipped. The transport web of these movements makes almost untraceable the exact routes followed by the materials, but we can make some assumptions. We assume that 1,539 MWp of modules have traveled some 10,000 km in an overseas route, with a domestic route of 300 km. by road or railways transportation to the port/airport in the country of origin and another 500 km. of road transportation from port/airport to the final destination site, including some transits through a logistic store.

In Spain, most of the installed PV plants in 2008 were made in a rush to arrive at the September 2008 deadline due to the regulations. We estimate that at least 5% of 1,500 MW were shipped by air from distant places, to save about one month in the delivery of material compared to transport by merchant ships. Approximately 1,500 MW of modules were sent from Germany to Spain and traveled 3,000 km from origin to destiny in the respective parks and all of them by road in containers. We assume that for Spanish modules there was 500 km of average transportation distance from factory to the sites. A 40 feet container has the capacity to carry 580 modules of 200 Wp each or 520 modules of 225 Wp. This makes approximately 116 kWp of modules per container. Therefore, on average, a total of about 30,000 containers of 40 feet (12 m length) have been shipped from different distances to provide the modules conforming to the 3,838 MWp of solar PV-installed plants in Spain up to 2009. This is 60,000 TEUs (20 foot equivalent units) in standard transport terms. Each TEU has a volume of approximate 360 cubic feet (39 $m^3$). Of this amount, we assume that 1,539 MWp have been shipped by sea. This represents 24,428 TEUs. A conventional container ship having 8,600 TEUs, has a power of 65 MW and a cruising speed of

**Table 6.8** National production of main parts of the solar PV parts in Spain

| Spain 2008 in MW | Ingots | Cells | Modules | Inverters | Trackers |
|---|---|---|---|---|---|
| Actual production capacity | n.a. | 195 | 498 | 741 | 414 |
| *Balance of import/exports* | | | | | |
| National production | n.a. | n.a. | 487 | 741 | 414 |
| Exports | n.a. | n.a. | 78 | 15 | 16 |
| Imports | n.a. | n.a. | 2,187 | 1,879 | 578 |

*Source*: ASIF Report July 2009. Pages 16 and 17

24.6 knots (46 km/h). We assume that these container ships have a return ratio of 1.5. About 3 loaded ships with about 1.5 empty return equivalent trips, will be sailing 15,000 km, lasting each 326 h full ahead. This represents an energy consumption of about 326 h×65 MW×4.5=88,000 MWh=95 GWh.

We subtract 5% of this energy expenses for transport by sea, to be shipped by air, as mentioned before, this resulting in a total energy expenditure by sea of 90 GWh. The 5% sent by air implies about 3,000 TEUs of solar modules shipped by air. The total volume occupied here being 3,000×360 cubic feet=about 1 million cubic feet or 108,000 m³ for the averaged assumed distance of 10,000 km. A typical cargo plane such as the Jumbo cargo plane 747-400F has a total load volume of 29 containers 10 feet long, respectively. Therefore, some 205 long-haul (10,000 km average) flights will be needed for this fast delivery transportation; this typical cargo plane is quite efficient in fuel consumption, so we assume a conservative approach. Its flight range is 1,679 nautical miles. We assume here return trips with only 30% of load occupancy, but with almost half the fuel consumption. So there will be 205×1.5=307 equivalent loaded long-haul (10,000 km) air trips of this type of average cargo plane. As the range of this plane is 4,445 nautical miles (8,230 km) and its consumption for this distance is about 50,000 gallons (190,000 l) of kerosene, the total consumption estimated for this type or air transportation will be 307 ×(10,000/8,230)×50,000= 19 million gallons of kerosene=72 million liters of kerosene. At a conversion rate of 1,000 l of kerosene=10.5 MWh; therefore, all the airfreight energy consumption would be in the range of 756 GWh.

And finally, we consider the road transport of containers as given in Table 6.10. We assume transport with the largest type of containers (40′) to optimize consumption (which is not always the case, as there are many accesses that allow only small trucks or even pickups) and use an average consumption for all type of national trucks in the USA of some 0.45 l of fuel per km (FERC) and that 70% of them return empty. So, we use a conservative 0.3 l/km for the combination of consumption of round trips of loaded/return unloaded. That, for this land transportation, is about 6 million liters of fuel. Using the conversion of 1,000 l=10.7 MWh, we obtain some 64 GWh for the energy expenses of the road transport segment of the 3,830 MWp of modules from their origins to their different destinies in Spain. We have ignored here the additional energy used in trans-shipping, transits, cranes and lifts in ports/airports, intermediate downloads and uploads in stores/warehouses, short distance

Table 6.9 Simplified scheme of main processes in the PV sector and main countries associated with their manufacture

| MW for Spain | Manufacturers/producers | | | | | System and serv. Energy | | |
| Country | Ingots | Wafers | Cells | Modules | BoS | Providers/EPC's | Sellers | O&M |
| Players | Handful | Tens | Tens | Few hundreds | Few hundreds | Many hundreds | Thousands | Hundreds |
| Japan | 2,500 | 2,830 | 2,830 | 400 | 1,539 | | | |
| USA | | | | 300 | | | | |
| China | 780 | | | 1,500 | | | | |
| S. Korea | | | | 100 | | | | |
| Taiwan | | | | 100 | | | | |
| Others | | | | 80 | | | | |
| Germany | 500 | 800 | 800 | 300 | 1,500 | | | |
| Spain | | 150 | 150 | 1,000 | 741 | 3,668 | | 3,668 |

**Table 6.10** Scheme of approximate km per road of containers transporting 3,830 MWp for the Spanish installed PV plants up to 2009

| Origin | MWp | TEUs | km by road | Total km × TEU | Total km × 40′ container |
|--------|-----|------|------------|----------------|--------------------------|
| Spain | 1,050 | 16,666 | 300 | 5,000,000 | 2,500,000 |
| Germany | 300 | 4,762 | 3,000 | 14,285,714 | 7,142,857 |
| Overseas | 2,480 | 39,365 | 500 | 19,682,540 | 9,841,270 |
| Total | 3,830 | 60,000 | 3,800 | 38,968,254 | 19,484,127 |

movements with pickups and lifts, etc. because of the difficulties in measuring all these movements. This gives a total, for the transportation energy expenses of finished modules as 88 GWh by sea, 756 GWh by air and 64 GWh by road, for a total of 908 GWh for the transportation of finished modules.

These shipments of finished PV products are far from being the only energy expenses in transportation. First, there are huge movements of raw materials to the grinding or crushing facilities, second, the treating of the minerals to obtain the specific ones; and third the transport to the melting plants or blast furnaces and from there to laminations, or cutting or performing, if not in the place and from there to the assembly lines. Considering that this is a globalized and specialized world, it may easily happen that many of these materials travel thousands of kilometers to end in the module assembly factory. We estimate (guess, really) that for these transportation expenses, some 500 GWh was needed for the materials, parts, and components of the installed PV park in Spain up to 2009.

Then there are inverters, trackers, and the rest of equipment (security, communications, control room, housings, low and medium voltage, monitoring, etc.) which represent between 40% and 50% of the module costs. The traveling costs certainly are lower than the modules, as these equipment and materials may be produced locally. So, we estimate (guess) some 500 GWh for the remaining equipment for all the installed PV base in Spain up to 2009.

There are also transportation movements of the technical personnel during the construction phase and after that for 25 years in the O&M like security patrols or O&M expenses. This would seem difficult to calculate, but we have experience with one plant of 1 MW (typical, but with a new technology that had to be especially cared for). For this plant at least 150 round trips of 400 km by road were made by the developers; some 300 trips by engineers with an average of 500 km in a mix of plane + rented car to fix SW and HW issues. And not less than 1,000 local 20 km round trips in pickups over 3 years to do routine maintenance jobs. Plus about 50 visits of other customers to see the plant, inspectors, and comptrollers and about 50 times 30 km trips of subcontracted cranes, tractors, repair and maintenance trucks, subcontracted people, auditors, quality control experts etc. This is 60,000 plus 150,000 plus 20,000 plus 1,500 or 231,500 km minimum. Assuming 0.15 l/km this would be 34,725 l. At 10.7 kWh/l this would be 372 MWh. Multiplying this value by the 1,400 GW of installed power in Spain gives 520 GWh. Thus, we guess another 500 GWh of energy expenses for personnel movements, making the total global approximation for transportation costs roughly 2,400 GWh. Even this ignores many hidden or disguised energy expenses in long haul trips or national air or road or train trips for businessmen to close deals, signing of contracts between distributors and manufacturers, of foreign investors and promoters coming to visit the sites, to supervise the progress of works, to visit banks to get credits, or institutions to get permits, etc.

In conclusion, we estimate a global transportation cost of 2,400 GWh for the transport expenses for the installed base in Spain up to 2009, as one-shot energy expense for the 25-year life cycle. Brought to present value it will represent $a_8 = 96$ GWh/year, equivalent to 1.9% of annual energy output.

The important thing, here, is not so much to undermine previous LCAs and EPBTs on this transportation aspects but to show that this type of energy is still *absolutely* dependent on a very complex, interlinked fossil fuel society, for which, nobody has seen a reasonable, functional replacement with the modern renewable energies.

**$a_9$ Premature Phase out of Unamortized Manufacturing and Other Equipment**

One of the most paradigmatic, yet contradictory, subjects in this analysis is the complacency of the industry in, on the one hand, announcing that the technology evolves so rapidly and improves the efficiency of the photovoltaic systems so rapidly and on the other, the obsession to disguise the necessary premature phase out of the previous manufacturing equipment with every new generation. The latter implies an obvious extra economic and, hence, energetic cost to the industry that should be accounted for.

Progress as we have seen before costs money and therefore costs energy. In less than one decade, we have observed the multiplication of technologies. Mono and poly-silicon (single-multi) were already popular but have changed substantially in size, wattage, and technical characteristics, such as output voltage etc., so that many of the PV installations have no possibility of replacing older modules when they fail, one by one, and they force owners to organize logistics to match the new replacements with the old equipment by throwing them away and then purchasing complete streams or arrays and connecting them all to the inverters. There have been a myriad of new technologies; some of them that may have required huge investments and have not finally succeeded as originally expected.

We have seen and heard promises and wonders about amorphous or micro-amorphous technologies, associated or not with thin film technologies of various types (nanotechnologies applied to photovoltaic modules, Cadmium Telluride (CdTe) cells, organic cells, flexible cells, manufactured in rolls that are deployed on roofs, slate modules, three or multi-junction cells, back-junction cells with 20–40% efficiency, photovoltaic graphene, organic with few atoms thickness on a flexible polymer, each promising to reach incredible efficiencies at negligible prices. It is beyond the scope of this study to enter in detail each of these technologies, promising always to improve the ratio of efficiency/cost at different moments of the last decade. This has not seen the light yet, but only in negligible volumes, as per the developments in Spain up to the date of publication of this book.

Manufacturers know that when they make an investment decision in a production line, a test lab, an acquisition of a R&D entrepreneur having patented a new technology, they have to plan for an economic pay back period of not less than 5 years while assuming that the new line will meet the load as per the theoretically promising growing market while increasing market shares. The reality is becoming less and less coincident with the plans. The causes may be multiple. Sometimes, it is because a new technology supersedes the old one in the market, before the latter has been amortized. Within the same technology (i.e., thin film), some particular version seems to be preferred to others, while the company may have just made the investment in a production line that will never be operated, often because the high efficiency modules are not worth the price.

In Europe, it has happened also that the same type of manufacturing technologies, when installed in China, have pushed the European companies out of the market, perhaps in the first or second year of the investment in Europe, because all other cost factors are better in China, even including customs duties and transport to import them into Europe. Brand new production lines remain idle in both Europe and China.

This may be a good business for the German, Swiss, Japanese, or US designers and providers of the very sophisticated manufacturing, testing, or automated assembling equipment, while also demonstrating that extreme competitiveness may be good for the market, while at the same time bad for energy saving purposes. The faster the new technologies and the automated production lines evolve, the more sophisticated they become and the shorter the life cycles and the higher the energy spent in them before they have to be phased out and scraped.

Of course, R&D is welcomed, when it makes energy sense, and while it leads to improvements in the real energy efficiency of a complete PV system (not only of a particular or specific type of cell), it is offset to some degree by the extra expenditures of premature equipment obsolescence on one side, or by another better technology arriving 1 year later on the other side. For example:

1. Expensive high theoretical efficiency (20%, for example) multi-junction cells that were designed for high concentration processes have become obsolete by new multi-junction cells of 40% theoretical efficiency, even before having proved competitive with conventional mono-poly-silicon modules. But they are still fighting to find a market niche.
2. The continuous progress in cutting silicon ingots into thinner and thinner cells, may leave machines with cutting saws able to produce 200 μm wafers obsolete much before the amortization period if 180 μm becomes a proven standard with better manufacturing costs.
3. The massive deployment of a given technology may become impractical due to specific laws and regulations.

It is certainly difficult to evaluate the energy expenses of the obsolescence provoked by the fast rolling/evolving technologies that sometimes die before they yield the foreseen advantages at global level. This may be especially the case where governments subsidize the manufacturers, as in China. But some hints may perhaps be extracted from the report of the Association of PV producers in Spain. The report mentions the "not used" or "spare" production capacity of the Spanish manufacturers, even in the booming year of 2008, where Spain was almost half of the world market for PV-installed plants. The idle or spare capacity is given in Table 6.11.

**Table 6.11** Percentage of not used capacity of the different PV parts of the PV systems of the national Spanish producers in 2009

|  | National production in MW | | | Total |
|---|---|---|---|---|
|  | Real production | Idle capacity | Idle capacity in % | |
| Cells | 195 | 65 | 25 | 260 |
| Modules | 498 | 393 | 44 | 891 |
| Inverters | 741 | 412 | 36 | 1,153 |
| Trackers | 414 | 327 | 44 | 741 |

*Source*: ASIF Report July 2009

We have to assume that the situation of lack of production and idle capacity of the Spanish producers was much worse the following year 2009 because during that time several other new entrants into the Spanish market increased the national output capacity with huge investments in manufacturing facilities.

We use a simplified procedure to assess the energy value of the premature phase-out of the unamortized manufacturing equipment. Assuming a theoretical 500 MWp yearly production capacity in modules and similar production capacities in inverters and trackers at an overall price (modules + inverters + trackers) of 3 Euros/Wp, this would mean a production capacity of 1,500 million euros/year. Assuming that machinery and equipment is acquired and amortized in 10 years and that it represents 50% of the total manufacturing costs, this represents investments of 750 million euros every 10 years on average, or 75 million euros per year in equipment replacement for all the Spanish solar PV industry.

If we assume a conservative obsolescence and phase out that affects 50% of the acquired PV equipment, this is an economic phase out cost of 37.5 million euros/year for the Spanish manufacturing equipment. This is 37.5 million euros, at a conversion rate of a typical engineering cost of 3,960 MWh/Million euro, represents ($a_9 = 148.4$ GWh), equal to 2.8% of the 5,069 GWh/year net energy (real yield) averaged output for the years 2009–2011.

## Operating Expenses

### $a_{10}$ Energy Costs Associated with Injection of Intermittent Loads: Pump Up Costs and/or Other Massive Storage Systems, When Applying

One of the most difficult to assess and elusive costs is that of the infrastructure and systems that would permit the intermittent modern renewable energies (namely wind and solar) to work in a national or international connected electric network without the need of fossil-fueled power plants for backup. This is normally not included at all in most of the LCAs and EPBTs analyses that argue that this is not the ultimate objective of the solar PV plants. However, if solar PV systems are supposed to be the long-term replacement of fossil fuels into the future, we need to consider them as such in this EROI study. Thus, we must consider backup or storage systems.

Electricity cannot be stored as such except as trivial (from a PV global perspective) amounts in batteries. But it can be stored as elevated water, in "pump up" systems between reservoirs which have a height differential. Pumps can become generators and the converse. In Spain, there are presently some 2,750 MW of installed power to pump up water from a lower to an upper reservoir, which then can be released during periods of the maximum need. Some 30% less electricity is regenerated than is used to elevate the water because of necessary losses due to the second law of thermodynamics and various systems inefficacies. This storage and production system is equivalent to 2.8% of all the installed power, and does not generate energy, but stores it for a later use. This installed capacity in 2009 pumped and then consumed 3,703 GWh, which represents a consumption of 1.34% of the total generated energy in Spain. As the estimated average losses are around 30%, we calculate that the losses in this system are about 1,111 GWh/year, equal to about 20% of solar electricity generated.

The reality is that most of this storage capacity was created in the 1970s and 1980s of the last century, to back up the eight existing nuclear power plants as they cannot be regulated to follow the demand curve variations at the speed they happen, that is the day/night peak/valley cycles of the demand. These pump up systems are able to "digest" or "swallow" up to 2,750 MW from the nuclear power plants, whose total national capacity is 7,716 MW. Thus the pump up storage needs for the nuclear power plants amount to some 35% of the total nuclear installed power, a significant percentage. This is to cover only the day/night (peak/valley) 24 h cycle variations of the demand, as the nuclear power output is nearly constant.

We know that this pump up capacity should be somehow preserved for the nuclear power-installed plants and that they could be used only partially for the purposes of stabilizing the network from short-term intermittencies (fast ups and downs of sometimes up to 2 GW of power in several minutes) provoked by the modern renewable sources, mainly, wind energy. They are less useful for solar whose main changes are day to night.

In several forums, responsible officials of REE have mentioned the need to increase Spain's capacity to some 6,577 MW of installed pump up power (IDAE 2011). This extra pump up power is supposed to be needed to cover the programmed power developments to 2020. By this year, according to the plans of the Ministry of Industry, Tourism and Commerce drafted in a document of June 11, 2010, they expect to need this pump up storage, basically to manage and stabilize the electric network in 2020 (IDAE 2011), with the following penetrations of the modern renewable sources, which are intermittent. Table 6.12 gives the expected energy mix for 2020.

Consequently, a system pretending to stabilize the electric network at the level of about 50% of their national demand renewable energies, from which about one third are intermittent renewables (wind plus solar), will need, as per the Ministry estimates, some 6,577 MW, two and a half times the present amount.

Although we recommend that this issue be investigated further, as it is a key issue if large deployments are foreseen to replace fossil fuels, we do not consider here the amount of embodied energy required to make the infrastructure required to stabilize the network from the intermittencies of solar PV, as it is difficult to make an educated guess while projecting into the future. This exercise is only to show the

**Table 6.12** National electric balance in Spain for 2020 according to the draft of the National Action Plan for Renewable Energies of the Ministry of Industry, Tourism and Commerce. June 2010

| National electric balance in year 2020 (PANER draft) | GWh |
|---|---|
| Coal | 33,500 |
| Nuclear | 55,600 |
| Natural gas | 141,741 |
| Oil products | 8,721 |
| Renewable energies | 152,835 |
| Pump up hydro | 8,023 |
| Gross production | 400,420 |
| Consumption in generation | 8,878 |
| Net production | 391,542 |
| Pump up consumption | 11,462 |
| Balance of int'l exchanges | −25,199 |
| Demand (in bars) | 354,882 |
| Consumption in transformers | 5,800 |
| Losses in transport and distribution | 31,138 |
| Final demand of electricity | 317,944 |

| Installed power of intermittent renewables by 2020 | GWh | MW |
|---|---|---|
| Wind onshore | 70,502 | 35,000 |
| Wind offshore | 7,753 | 3,000 |
| Solar PV | 14,316 | 13,445 |
| CSP | 15,353 | 5,079 |
| Total wind + solar | 107,924 | |

*Note*: Data in negative means exports of electricity

complexities of an electric network based mainly in renewable energies produced in intermittent conditions. They are forcing the network managers to forecast with backup power, the more as the renewable penetration goes higher. The energy investment and operational costs will be large and need to be calculated.

Therefore, $a_{10} = 0$ (for now, but could be large).

## Energy Estimates from Monetary Values

### *Financial Services*

#### $a_{11}$ Insurance

In order to be financed, all the PV plants need to have an insurance policy, covering all the most common and known risks: equipment stealing (the most common incident, for modules and copper cables), vandalism, and weather or meteorological incidences, such as windstorms, lightning, and hailstorms. Two main types of insurance are normally contracted. The Civil Responsibility, to provide coverage for provisions under civil law, usually in front of third parties, may be affected by the plant (electrocutions, fall of evacuation lines, accidents of third party labor inside the plant, etc.) and a full coverage insurance policy. The usual costs are around 1,000 €/MWp/year for Civil Responsibility. The other is the full risk insurance at a cost of about 10,000 €/MWp/year. We do not consider the degree of liability exclusions, that is when incidents or accidents occur and the additional economic (and energetic) expenses of many of these exclusions when the claims are presented, as many of the details of conflicts among customer and insurance companies in this respect remain undisclosed.

However, we are aware that some incidents have problems with receiving payments due from accidents, for various reasons that are difficult to attribute responsibility. For example in some cases equipment stealing is not covered if the alarms were not properly maintained or programmed. In others, the responsibility for destruction of modules due to hail- or windstorms may fall in a sort of nobody's land between manufacturers and constructors on the one side and the insurance company on the other, each variously claiming the damage was as a consequence of a faulty installation or maintenance or due to forces of nature beyond the contracted and insured specs.

Focusing then on the economic costs of the insurance policies, we estimate 7,500 €//Mwp/year of insurance costs, times 4,000 MWp = 30 million euros/year. By applying the Economic to Energetic equivalences of 660 MWh/Million euros for this activity, considered to be of the type business and financial, being that one third of the national energy intensity mean, we derive energy use equivalent to 0.4% of the energy output ($a_{11}$ = 19.87 GWh/year).

## $a_{12}$ Fairs, Exhibitions, Promotions, Conferences, etc.

One of the activities that also consume energy is the attendance and organization of the numerous fairs, conferences, and promotion and marketing activities. Some of them are difficult to allocate to a given, specific country because they involve international firms and may have interests in promoting products and services other than the ones intended for the hosting country. We give a few examples to illustrate briefly how much activity is involved around the solar photovoltaic world.

GENERA is the annual Trade Fair of Energy and Environment that took place in Spain in 2010, despite the economic crisis (in 2008 it was much larger). It had the following characteristics:

- 415 participants and 662 offering companies. 37% foreign companies
- 30 countries represented
- 28,984 attendants from 62 countries. 7.2% of visitors coming from abroad (Portugal, Germany, France, Italy, United States, China)
- 3 days' exposition

- Energy sectors: biomass, cogeneration, fossil fuels, wind, geothermal, hydraulic, hydrogen, marine, solar PV, and solar thermal. Solar PV was some 9% of the interest in trade visitors
- Some 35,000 m$^2$ of pavilions with 50% occupancy
- International and national return flights; national and regional long-range land transportation for visitors and exhibitors
- Cargo shipments of equipment (air freight and/ or vessel and/or truck) for materials
- Fitters and decorators
- Car renting, taxis, parking
- Electricity consumption, communication, etc.
- Hotels
- Printing catalogues and brochures

We have estimated the annual overall budget/expenses involving photovoltaic activities for this Trade Fair at around 10 million euros/year. Sometimes, there are congresses that are devoted not only to Spain and the Spanish environment, but that also represent expenses for Spanish PV and their suppliers such as the Solar Mediterranean Plan that took place in Valencia in May 11 and 12, 2010, with the attendance of 800 visitors from countries of all the Mediterranean basin. A few months later, a conference on solar PV technology, the 25th European Photovoltaic Solar Energy Conference and Exhibition was held in Valencia, on September 6th through 10, 2010, this time uniting three major conferences, the 25th European Photovoltaic Solar Energy Conference and Exhibition, the 36th US IEEE Photovoltaic Specialists Conference, and the 20th Asia/Pacific PV Science and Engineering Conference. They were combined, probably due to the world financial crisis.

Other related solar PV activities are located outside Spain but involve Spanish companies wanting to promote their technology and skills abroad, such as the Intersolar, another fair taking place in Germany. These activities are almost a continuum, and we have heard some manufacturers, EPCs, engineering Cos., consultants, and the like expressing tiredness for being forced to spend time, money, and energy to continuously show up in the market. After all, it seems that solar PV energy is green and sustainable but moves totally in a (very unsustainable and fossil fuel intensive) business as usual form with these activities.

Solar Decathlon is an international contest to show homes powered with solar PV and thermal energy and prepared for energy efficiency and energy savings. In April 2010, the conference took place in Madrid and it has been agreed between the USA and Spanish Administrations to hold this annual event in alternate years in both the USA and Spain. In this year, 5 out of the 17 solar PV home projects were originated in Spain, mostly from subsidized academic or university sponsors.

Sometimes, the activities involving the solar PV world are not directly related with the manufacturing of modules, but have to be considered. In the last 2 years, a lot of conferences, symposiums, seminars, etc., have been imparted on the so-called Intelligent (Electric) Networks, which are very much related and interlinked with the modern solar and wind technologies and their connection to the national grids.

Clearly, the components are intelligent but can we say the same about the human-designed systems they are supposed to manage?

We have included here all the activities and expenses to obtain certifications, sometimes expensive, required to sell any PV product. They are an imposed toll required to obtain the theoretical guarantees. For example, the following specs usually apply for modules and/or inverters:

- Norms IEC 61215 and IEC 61730 (others pending for HVPV systems).
- IECEE's certification of compliance stamps by appointed and specific labs.
- CE marking in compliance with 73/23/CEE normative for low tension; 93/68/ CEE directive and the 89/336/CEE directive on electromagnetic compatibility.
- 73/23/CEE normative on security essentials to electric appliances.
- 93/68/CEE directive, modifying the 73/23 CEE normative.
- EN-50178 and 61000-6-4 normative and security related issues.

If we add in the usual marketing expenses, brochures, or subscriptions to the growing number of associations in the sector (i.e., Secartys, ASIF, AEF, APPA, AETIC, among others), we may conclude that 40 million euros/year expenses is a very conservatively low amount of money invested in the PV sector in Spain for this factor. The energy equivalent to the 40 million euros spent in fairs, exhibitions and similar activities. By applying the business and financial economic to energetic equivalence (one third of the national mean). Therefore ($a_{12} = 26.4$ GWh/year), about 0.5% of the energy output.

## $a_{13}$ Administration Expenses

There were 57,900 individual registered PV plants in feed-in mode in the Spanish electric grid. It is quite common that plants have a size of 100 kW or less. The PV plants can be assigned to an individual or to a legal entity, usually a limited liability company, normally a Special Purpose Vehicle or Entity (SPV/SPE). The latter is pre-ferred due to a much easier transfer by just acquiring the shares, while the individual needs to request a change of the name of the holder, which is always more complex. In either case, individuals or companies may or may not decide to hire an agent to admin-ister invoices, incomes, and expenses and to control VAT and other taxes, in correspon-dence with the Ministries or Communities and Municipalities, etc. Most of the PV plant owners, and especially the big investors, do hire an agent and sign these types of administration contracts. The usual fees are in the range of 1,500 €/100 kWn/year, including extra expenses covering the formal annual meetings, auditing of books, etc.

We are not including here some other SPV's/SPE's created to exert control on a number of individual SPV's within a PV farm or group of PV plants.

Therefore, we assume that 60% of the PV plant owners do contract an agent for administration purposes. Thus, the total administration expenses for the Spanish PV total installed plants amount to some 34,740 individual installations with administra-tive agents contracted × 1,500 €/kWn/year = 52 million euros/year, so that by apply-ing the economic to energetic equivalences considered here to be of the type business/ financial, we derive here ($a_{13} = 34.3$ GWh/year), about 0.7% of the energy output.

## $a_{14}$ Municipality Taxes, Duties, Levies, and Taxes on Production

The urbanism taxes for solar PV plants have been evolving over time in Spain. At the beginning, in the booming real estate period of 2004–2007, municipalities in Spain were living in a booming time of real estate construction and they were receiving most of their income from the taxation of these sources or the revaluation of municipal lands for this real estate business. During this time, they asked those applying for urbanism and work license permits for a 2% of the investments for the PV plants for the urbanism license and 2.6% on the civil works amount for the work license. That was a negligible amount that represented about some 20,000 €/MWn on average, when the PV plants were at the price level of 4–5 M€/MWn in fixed mode and 6–8 M€/MWn for two tracker axes. The civil works represented a minimum portion of the total value of the plant on a turnkey basis.

Money collected from taxes is primarily spent on infrastructure projects such as bricks and mortar and activities, such as road repair. To calculate the energy associated with taxation, we assume that the overall (turnkey) price per MWp was about 5 M€. Since in 2009 the total PV-installed capacity in Spain was about 3,668 MWp, we estimate the total investments for the PV national installed plants in 2009 at about 18,340 M€. In 2010 and 2011, new PV installations took place with an increment of 700 MWp more but already at lower prices per MWp installed (we assume here at 3.5 €/MWp). If municipal taxes are applied to civil works only, and they are 0.4% of the total invested price for the PV plants, this amounts to a total of 20,000 €/MWp at the assumed price of 5 M€/MWp.

Then, in coincidence with the bursting of the real estate bubble, and in parallel with the diversion of most of the huge financial capacities of the real estate corporations relative to the world of the renewable industry, the municipalities joined efforts to search for a new source of funding to keep their infrastructures functioning and started to tax not only on the civil works. Immediately, they eyed the burgeoning PV industry, which of course was not interested in being taxed more as the PV plants were already paying from 2% to 4% of the *total* project costs as taxes. Finally in May of 2010, the Spanish Supreme Court settled in favor of the municipalities, giving them the right to tax civil works and also machinery, equipment and installations of solar PV plants. The municipal vultures circled the fat carcasses (PV plants), which received subsidies that ultimately came from Madrid. The amount of taxes varied from community to community and from one municipality to another, but it was clearly a coordinated action and common willingness of the municipalities to increase their taxes to the maximum extent possible. This coincided with the first wave of rumors that premium tariffs were enriching the promoters and also with the plummeting of the real estate business.

We take a conservative figure that half of the 3,668 MWp installed PV power up to 2009 was taxed at 20,000 €/MWp and the other half at 2% of the turnkey price at 5 Meuros/MWp. And then, finally, that 90% of the 700 MWp installed in 2010 and 2011 were at 2% based on 3.5 €/MWp.

Therefore:

- $(3{,}668 \text{ MWp}/2) \times 20{,}000 \text{ €/MWp} = 37$ million €

- $(3{,}668 \text{ MWp}/2) \times 5 \text{ €/MWp} \times 2\% = 183$ million €
- $700 \text{ MWp} \times 3.5 \text{ €/MWp} \times 2\% = 49$ million €

This suggests that the total cost of municipal taxes were on the order of 269 million euros.

Of course, these estimated taxes are only one shot at the beginning of the works and are very conservative. Divided by the 25 years of the theoretical life cycle we obtain some 10.7 million euros/year. Some other taxes apply on a yearly basis. They are called "Bienes Inmuebles de Caracter Especial" (BICE) and normally are additional and complementary to the "Bienes nmuebles Rústicos," a taxation on the land in rural areas, where virtually all the installations are placed. These latter taxes are lower than if the plants are installed in an industrial or urban area (on the ground), where taxations on the land are heavier.

We assume the lowest taxation as if all installations were in rural areas. The estimated amount for these combined taxations is on the order of 2,000 €/MWn/year. This makes a total of 7 million euros/year.

A new tax in 2010 was imposed on the electricity producers including the wind and solar PV producers. It represents 0.5 cents of Euro per kWh. Assuming some 6,624 GWh/year produced and invoiced at low tension meters in the averaged triannual 2009–2011, this represents a tax of 3.3 million euros/year, which were not originally considered by most of the PV promoters and producers. Therefore, the total of 21 million euros considered for the different taxes in 2009, (which then require energy when used—which we assume to be of the type business/financial) represents some 0.3% of the energy output ($a_{14} = 14$ GWh/year).

It is worth noting that as a consequence of the economic and financial crisis in Spain 2008–present, the government has decided to impose a new tax on generation for all PV plants consisting in 6% of the total revenue as from January 2013 onwards. This has followed an intense debate in the Spanish media in the second quarter of 2012 which is flooded with articles, comments and editorials talking on the necessity to collect taxes from renewables or to cut the premium tariffs, something that is strongly opposed by the PV investors because it could wipe out their profits.

## $a_{15}$ Cost of Long-Term Rents or Ownership of Land for Solar PV Plants

Spain has installed almost all of its photovoltaic capacity on the ground (Table 6.13), which is a very different approach from that of Germany, which uses rooftop PV installations to a much greater extent.

The process of acquiring land, usually in rural areas (urban areas were too expensive previous to the bursting of the real state bubble and even today still are) became increasingly complex and more expensive. The complexity resided in the time and administrative efforts to first qualify the land for PV plants and then to change the classification type. The regional administrations in Spain tend to promote more the use of dry lands and are quite opposed to give permits to lands qualified as irrigated lands, which are naturally the most expensive. However, some regional administrations accepted installations with big two axis trackers on irrigated lands if they

**Table 6.13** Split of PV installations in Spain by the type of surface employed

| | |
|---|---|
| Rooftop installations in Spain | 2.2% |
| On the ground installations in Spain | 97.8% |

*Source*: ASIF Report 2009

provided proof that the use of the land would remain mainly for agricultural or cattle ranching purposes. Many environmentally protected areas in Spain are forbidden for this use: natural parks, other protected areas and the *Zonas de Especial Protección de Aves* (ZEPA or Special Protection Areas for Birds, for its name in English) or *Lugares de Interés Comunitario* (LIC, or EU Places of Interest, for its name in English).

There are also some regional rules that may vary from one autonomous community to another on how to keep minimum distances to roads or motorways. The selection process normally starts by considering the price of the land and its suitability; the lack of trees or topographic conditions that may project some shadow on the modules, access to heavy trucks, cranes, and vehicles and the overall distance to an access and connection point that is able to evacuate all the projected energy of the PV plant into the grid. See next factor in this respect.

It is assumed as a standard in Spain that 1 MWp occupies some 3 hectares (ha) of land (lanes, housings and minimization of shadowing included) for fixed PV plants and about 6 ha for PV plants with two axis trackers. Therefore, the space occupied by the 3,668 MWp of PV power installed at the end of 2009 totals of 14,360 ha (Table 6.14).

It is difficult to analyze the total PV installations in terms of costs of ownership and renting of the land. Originally, the trend was for the first developers to buy the property at prices varying from 2,500 €/ha in dry lands up to 12,000 €/ha in irrigated lands, if legal permissions for the latter were granted. But prices skyrocketed very rapidly when farmers learned about the high margins of PV plant developers. The main investors began to undertake long-term renting leases, since it was better for their financial purposes and saved taxes on their books. The renting contracts are usually made for a minimum of 25 years, with unilateral extension clauses, for another 25 years in favor of the lessee. These contracts usually foresee an initial renting price per ha or total renting price for the required plot of land, and include an escalation formula linked to the Consumer Price Index (CPI).

A conservative renting price of lands for PV plants in the mature stage of development would be around 1,000 Euros/ha/year, although there were commercial contracts in the booming phase of the PV installations in 2007 and up to mid-2008, where prices reached up to 1,500 €/ha/year. The PV companies that have opted to own the lands had to pay irrigated land prices for dry lands, which averaged 12,000 €/ha/year or even up to 25,000 €/ha/year. We assume that 75% of the occupied lands by PV plants are in long-term rental contracts and the remaining 25% under ownership. We assume 1,000 Euros/ha/year for those lands rented and 17,000 €/ha for lands bought, which amounts to 680 €/ha/year for those acquired in ownership. We assume that land price inflation will not take effect during all the life cycle of the plants. The total cost of land will be then 14,360 ha × 75% × 1,000 €/ha/

**Table 6.14** Analysis of space occupied by the different types of PV plants in Spain

| Type of PV plant | % | Installed power in MWp | ha/MWp | Total hectares occupied by PV plants in Spain 2009 |
|---|---|---|---|---|
| Fixed plants | 63% | 2,311 | 3 | 6,933 |
| One-axis trackers | 13% | 477 | 4.5 | 2,146 |
| Two-axis trackers | 24% | 880 | 6 | 5,282 |
| Total | 100% | 3,668 | | 14,360 |

*Sources*: ASIF Report 2009 and own elaboration

year $+ 14{,}360 \times 25\% \times 680€/\text{ha/year} = 13.2$ M€/year, so the energy cost associated with land acquisition (using the business/financial economic to energy equivalence) is 0.2% of the energy output ($a_{15} = 8.7$ GWh/year).

## $a_{16}$ Circumstantial Labor and Associated Economic/Energy Costs (Under V. Financial): Consultants, Notary Publics, Public Register, Civil Servants/ Public Officers, Engineering Colleges, Legal Firms, etc.

Some additional business/labor costs usually are not included. All of this business activity requires energy to build and operate the consultant's or other person's building, for the numerous trips by automobile to fill out forms, discuss terms, and so on; to make the computers, paper and other business supplies, and so on; as well as to give meaning to the salaries. All of these (prorated for the PV portion) are legitimate and necessary energy costs of building and operating a PV facility. We assume the national mean energy intensity per Euro would apply.

Consultants

Consulting services are becoming more and more necessary as the banks force PV developers to present complete business plans. These plans give the banks the confidence they need to finance their PV plants with leverages of up to 80% of the total turnkey project prices. Reputable consulting firms in Spain were offering this confidence, ensuring both promoters and banks that every document of a given PV field file was adequate and accurate.

We documented over 70 kilos of physical paper for a 1 MW project. The pile of legal documents is essential to specify or give evidence of:

- Property ownership or long term rental of the plot/plots of land
- Evidence of project execution (five copies) duly stamped by the corresponding Engineering. College, including the modifications and/or retrofits of the original
- Adequate point of access and connection to the electric network, issued by the electric power utility and evidence of having paid the feasibility study fees
- Environmental study presented and approved by the corresponding authority, without restrictions placed on the development

- Urban qualification of the plot of land for the use of electric generation under special regime (sometimes it takes months) issued by the corresponding autonomic authority
- Urban license issued and taxes paid
- Works license issued by the municipality, and taxes paid, including both the provisional and definitive forms
- Official Register of the PV plant in the autonomous region (REPE) per individual plant within the park or PV farm
- Official Register of the Plant at national level (RIPRE or RAIPRE) for each individual PV plant
- Opening and activity license issued
- Contracts for electricity supply with the electric power utility
- Contract with the agent representative
- Statement of connection (acta de puesta en servicio o puesta en marcha) all of them with signatures and stamps from PV installers, including evacuation lines
- Rights of ways with third parties, if any, duly legalized in notary public
- Certificates of compliance, guarantees, supplies, O&M contracts, etc.
- Contracts for security and insurance policies
- Multiple invoices.

To give a green light, these consultants have fees ranging from 30 to 60,000 € per due diligence (i.e., a legal term indicating that the operation has been carried out as endorsed or backed by a certified public accountant for each plant). The big investors acquiring the bulk of the PV parks are focusing on plants of 2 MWn or larger. We assumed that 70% of the PV-installed base has adopted these procedures and has spent 10,000 € per MWn in consultancy. This represent an economic expense of $3,668 \times 70\% \times 10,000 = 25$ million euros in consultancies per MWn.


Notary Publics

These are necessary to sign contracts for credits and leases, and to grant project financing. They are also required to register the SPV companies created to hold the individual PV plants, for transmissibility purposes. Since in most of the cases the legal entities are the individual PV plants of up to 100 kW, we assumed that some 25,000 deeds have been issued by notary publics for Spain at the lowest legal level of a limited liability company (3,006 €) plus the same number of deeds for the credits/leasing. In this case, we include the costs of inscribing the deed in the Mercantile Register. Notary publics are also required to issue the deeds when the land is purchased or rented by the land developer. In this case, the rights of way need to be legally registered to avoid conflicts. For this purpose, we assume about 20,000 public deeds for all of Spain. We estimated the costs of notary publics as follows:

1. 25,000 deeds for credits/leasings for all individual PV plants.

2. 25,000 deeds to constitute the limited liability companies for each individual PV plant (SPV).
3. 20,000 deeds to inscribe the plot of purchased/long-term rented lands in the Property Register under the name of a given SPV Limited Liability Co. and the corresponding rights of way, if any.

We did not consider the second and subsequent legal renditions of PV plants for the 2007–2008 period, which were frequently changing hands due to the new entrants in this very active market in Spain. The buy/sell movements may have multiplied several times the inscription needs and costs (economic, ergo energetic). A lump sum evaluation for the above activities would be in the range of 100 million euros.


## Social Costs to Develop Photovoltaics or Any Other Engineered Facility

### Engineering Colleges

There is a myriad of related and specific activities around the installed solar PV base that most likely are not accounted for, when assessing direct and indirect labor used to develop the Spanish solar industry. These are costs that are paid for from the general economy, or by the parents of the students or even the students themselves that must be met from the general, mostly fossil-fueled economy. For example, each and every PV project has to be signed by an authorized person (usually an industrial or civil engineer or telecom engineer), that is a member of the official certifying agency. This represent a cost that is normally included in a turnkey project, but sometimes the cost of the fees that the College of Engineers takes to endorse a project are not included. These projects are usually undertaken for a combined group of individual PV plants; that is, one per PV park or farm. We assume 3,000 projects endorsed by the colleges in the different provinces at about 6,000 € per project for grand total of 18 million euros. Of course, there are energy costs involved in training all the engineers, but we will not go there in this EROI analysis.


### Legal Costs

There were a growing number of trials in 2009, 2010 and 2011 related to disputes involving PV plants. The two main groups involved were promoters being sued by municipalities, asking for higher taxes for civil works or urban permits than those originally forecasted. Also, there were conflicts between the big financial associations and the government because of the continuous changes in regulations, which PV developers believed violated the basic principles of the rule of law—that is that a contract is based on the regulations in place at the time that the contract is signed. We assume that these legal firms charged some 20,000 € per each legal conflict they represent, either against manufacturers, regional administrations or other conflicts per MWn. The price per MW may be lower if the plants are larger or if the promoters syndicate a common claim. Since it is still a very uncertain figure and premature to calculate throughout the 25 years period, we do not assess or value this activity.

Summarizing economic costs for consultants, notary publics, engineering colleges and legal costs, we obtain 143 million euros. These are generally one-time expenses, usually in the installation of the PV plants, although there are some recurrent expenses in some of them that are difficult to assess. This is an energy equivalent to 247 GWh, but for the whole life cycle of 25 years, the annual costs would be 9.9 GWh/year.

### Civil Servants/Public Officers

As renewable energies began to develop in Spain, a number of civil servants and public officers at the two highest levels of the Spanish Administration had to be devoted full time to take over the administrative legal tasks related to these projects. These jobs were created for technical and electric generation issues related to the PV plants at the national level, mainly, in the Ministry of Industry and related specific dependences including the ministries of Tourism and Commerce and the Ministry of Environment, Rural and Marine for permits and qualifications. There are new positions created in the Ministry of Finance because renewable energies are under a "special regime" with some exemptions and special duties to the government. At the autonomic level (equivalent to the State level in the United States), there were additional specific jobs created, since the activities carried out at this level were much more numerous. This is a labor force obviously not included in our earlier assessment of Spanish direct and indirect labor. But it is obvious that energy is consumed directly and indirectly in the operation of these sectors. Normally, these are jobs that may have existed in other related energy departments. Thus real and comprehensive energy cost assessments would have to include this energy costs for each source of energy.

We understand that calculating the number of these civil servants is a difficult task, as is calculating the direct and indirect labor associated with the Central or Regional Ministry or agency, department or section of Spain. And within it, it is even more difficult to ascertain how many of the direct or indirect special regime public officers/ civil servants are dedicated to solar (PV, CSP or thermal) versus fossil fuels, or what the net number is given that they are not regulating some other source of energy at the same time. We assume a conservative 4 persons/year on average per autonomous region as direct, full-time labor devoted to the solar PV activities and 20 for the entire central administration in Madrid. The total pro-rated energy expenses of civil servants in public officers is then 88 workers times 86 MWh/worker=7.6 GWh/year.

### Electric Utilities' Workers, Electric Regulators' Specialists, IDAE, University Departments and Solar PV Associations

We debated on how to include the workers of the main electric power utilities of Spain (Iberdrola, Endesa, Unión Fenosa) and a few minor ones. It is clear that these utilities have had to increase their staff to attend to the wave of solar PV feed-in installations, especially given the intermittency of the solar operations. In both cases, this is an extra dedication to economic (and hence energetic) expenditures. The same can be said about the Spanish electric regulator, REE a public company, and several institutions that have direct labor specifically dedicated to solar PV systems. One of those institutions is the Instituto para la Diversificación y Ahorro de Energía (IDAE), a sort of public advisor to the Ministry of Industry and others in energy matters, that has a division specialized in solar PV plants and a proportional part of its management and services staff dedicated to solar PV. Many universities in Spain have departments with professors and researchers working part or full time on solar PV activities. There are also associations defending the interests of the sector, with employees being part or full time employed. We assume a lump sum for these employees of 30 workers full time at the assumed 86 MWh/worker and year = 2.6 GWh/year.

We note these economic costs to make the readers realize that additional social activities are devoted to the organization and coordination of solar PV activities. Thus their economic/energy expenses must be attributed to the solar PV energy inputs. All these recurrent expenses, such as civil servants/public officers represent about 11.2 million euros/year, so that as per the Economic to Energy equivalences assumed as a business/financial related activity we derive ($a_{16}$ = 22 GWh/year), about 0.4% of the energy output.

## $a_{17}$ Agent Representative or Market Agent

Since September 1 2009, the provisions established in the 662/07 Royal Decree forced all the PV plants with over 15 kW of installed power to sell their energy through "production offers" to conventional utility companies. They can do this either through the so-called "tariff regime," a market agent or the electric operator to which they are connected to the grid. The latter will automatically deduct 0.5 cents/kWh. Starting in November of 2009, the PV installations were automatically taxed 1 cent/kWh. However, virtually all the PV plants were under the "regulated tariff" regime (premium tariffs), so instead of taking an automatic deduction based on electricity production they had to present to the Electric Market Operator (OMEL) a daily forecast of probable production. PV plants can be penalized for substantially inaccurate forecasts (usually any deviation of +/− 5% on forecasted production). The rationale behind this requirement is to adequately program the necessary matching of the production to the demand curve at the national level. When the energy comes from combined cycle gas power plants or hydro they can be easily programmed to increase or decrease production, but with the advent of a significant portion of intermittent production from renewable systems such as wind and solar PV, the managing of these systems becomes more difficult.

The need for this continuous energy output forecast would have forced the owners in an individual solar PV plant to employ a person to be exclusively devoted to report this forecasts to the OMEL. Therefore, some private companies surged in the Spanish market offering to take over this responsibility. OMEL allowed them to forecast nationally and globally on all the solar, wind or conventional production plants, so that the energy production could be predicted with much more accuracy. In fact, there are now very powerful tools to predict electrical generations from solar PV or wind parks, hours in advance for entire regions. These tools include thousands of sensors gathering climatic data in the Iberian Peninsula and surroundings, connected to expert systems making the global prediction. These companies were able to sign contracts with solar producers in Spain at a fraction of the price (0.5–cents/kWh), and they also assumed the responsibility to pay the penalties for possible failures in the forecasts.

This is another economic cost that was not predicted by PV promoters, who installed their plants in the 2004–2008 years. We are aware of several solar PV producers that are still being automatically served by the Distributors of Last Resort (CUR) and are still paying 1 cent/kWh or 10 €/MWh due to their poor management. We assume, here, that virtually all PV plants have selected a private market agent. Although their rates vary, a common rate of these companies has been 1.5 €/MWh sold. Thus, the total economic/energetic cost of having a market agent for solar PV producers in 2009, when 6,074 GWh were produced, was 9.1 million euros/year, so that as per the Economic to Energy equivalences assumed this activity being of the type business/financial, we derive ($a_{17} = 6$ GWh), about 0.1% of the energy output.

### $a_{18}$ Equipment Stealing and Vandalism

Times are tough for many in Spain and one of the aspects never considered in the LCAs and EPBTs is the stealing of modules and other parts of the PV plants, especially copper, which has a high price in the black markets. Sometimes, even the computer equipment in the control rooms has been taken. The equipment stolen has to be replaced, and thus the PV plant ends up costing more than originally foreseen.

The PV modules are easily smuggled as scrap shipped and resold in third world countries, or to individuals within Spain. These stolen modules are generally used for small individual applications, for example, the recharging of car and truck batteries during the daylight in order to have small electric lighting or appliances during the nights. The fact that originally most of the PV modules delivered conventional 12 and 24 VDC, and from 150 to 240 Wp eased this stealing. These stolen modules cannot be reused in registered PV plants because they are required to have the original receipts ready for inspection, which are careful inspections. Specialized bands and gangs have been able to break into solar PV plants through fences. In some instances, they are able to dismantle between 60 and 100 modules in half an hour to one night, depending on their skills and difficulties.

To avoid this vandalism, promoters and PV plant owners have sharpened their security and surveillance systems, thus diminishing the stealing but at the same time increasing the security costs. Insurance policy rates are sensitive to the number of incidents, the value and the effects on the energy production. Thieves focus more on small and remotely located PV plants, accessible only by rural roads, than on big plants which are usually better endowed with sophisticated security systems. Security procedures to avoid the stealing of modules include wiring them in series of strings, so the dismantling of one of them breaks the circuit and acts as a tamper shooting the alarm. If the gangs are professional, these types of procedures do not work. Other attempts by manufacturers and promoters to prevent stealing go in the direction of installing modules with no commercial voltage outputs, or modules configurations in larger sizes (>500 Wp/module), which make them less mobile. There are PV manufacturers installing modules to their structures with some special glue or by blind bolds and nuts, so the attempt to dismantle will destroy the object of desire.

Neither the owners nor the insurance nor security companies are eager to release figures, so it is difficult to make estimates of the economic damage to the PV plants affected. It is also difficult to estimate the energy invested in the new modules to replace the stolen ones. Perhaps, some of the stolen modules are providing energy in a third world home for some time. However, the possibilities of modules lasting for the full specs of 25 years life-cycle under elemental and unskilled applications seem low, modules may also find their final use in an off grid remote low level application. Spain, after all, is a first world country, with a very efficient and quite incorruptible rural police (Guardia Civil), and a very good public security infrastructure. We wonder how the stealing will affect PV plants installed in most of the third world countries, where state protection or security structures are corrupted, or in many cases almost inexistent.

We have estimates in Extremadura, where 6 out of 20 known PV plants have been broken into. In these instances, smugglers steal also portions of the existing equipment and materials, such as copper from reels or already installed, and steel or aluminum structures, which they end up reselling as scraps in black markets. And of course, PV modules were stolen, sometimes still in their packaging. Valuing these stolen equipment and material losses is a difficult task.

In Spain, we have not detected, up to now, significant acts of vandalism. We know of a case in Italy, where a PV plant's modules were carefully shot with firearms, for possible reasons of disagreement with local companies and the promoter/investors. We use an indirect method and assign one tenth of the revenues calculated for the insurance policies in factor $a_{10}$ above. Insurance companies work for profit and a cost of 10% on sales seems reasonable, stealing and vandalism being the most important risks to them. This is equal to 3 million euros/year, so that at a conversion rate of a typical engineering cost of 3,960 MWh/Million euros or two times the energy intensity national mean, we derive ($a_{18} = 11.9$ GWh/year), about 0.2% of the energy output in 2009. We calculate this factor not so much for its energetic costs to Spain today, but because we believe that the social stability in most of the world is much lower than in Europe or the United States and in a potentially more chaotic future society, these risks may increase substantially (specially in very scattered and disperse premises) in an order of magnitude, both in the costs of the policies (energy invested) and in increasing losses of revenue (energy returned).

## $a_{19}$ Communications, Remote Control and Management

Communications in the PV plants are required by law to have digital meters remotely accessed to the electric power utilities to whose electric network the plants are connected. They are responsible for measuring the energy outputs, which have to be reported to the CNE. There is another compulsory communications system (UHF) that gives the connected power utility the ultimate authority to disconnect the PV plant from the grid for management, security or maintenance purposes. These expenses form part of the PV turnkey projects, and they are included in the turnkey prices. Many PV plants have access to data (and sometimes management control possibilities) to analyze the status of the PV plants, the meteorological conditions, the irradiances measured on the horizontal plane and global, and the online performance of the inverters.

One advantage of Spain is that virtually every plot of land has GSM, GPRS or UMTS (3 G) connections so that the PV plant can communicate with any place in Spain. In some cases, there are satellite communications or point-to-point digital low-capacity microwave systems that link the PV plant with the nearest location where the telecommunication fix (ground) network is available (usually the nearest town or village), of course, in this case, wherever there is line of sight between the two points. These technologies sometimes necessitate the installation of towers or masts at one or both ends. This point-to-point system then enters into a 3–10 MB/s ADSL copper-wired telephone line into the national telecom system. Fortunately for Spain, the ADSL penetration reaches not only cities, but almost any remote location of the country.

We make the following assumptions to calculate the energy cost of these communication systems. For large-sized PV plants of >5 MW, we assume a full communication system via satellite at 3,000 €/year connection fee per PV plant, plus 1,000 €/year for each MW for GSM/GPRS/UMTS connections, with data contracts for the low-tension meters. For medium size PV plants of 2–5 MW average, we assume only GSM/GPRS/UMTS connections for monitoring of the plant and security at 1,000 €/year per each MW. For small size PV plants of <2 MW, we assume 80% with only GSM/GPRS/UMTS connections and 20% with ADSL. The cost of the first group is around 1,500 €/year/MW for installation, and some 600 €/year for an ADSL line per PV plant.

For the 3,908 MWp of installed PV power of the size and distribution given in Table 6.15, the total estimated economic costs for 2009 are 4.3 million euro/year, so that at a conversion rate of a typical engineering cost of 3,960 MWh/Million euros, or two times the energy intensity than the national mean, this represents ($a_{19} = 17$ GWh/year), about 0.33% of total energy output.

## $a_{20}$ Pre-inscription, Inscription, Registration Bonds and Fees

In Spain, it was compulsory from 2009 onwards (until the government froze the new installations and pre-registration requirements) to deposit a bond of 500,000 €/MWn to have the right to install and commission PV plants. These obligatory bonds started as a measure to avoid speculation with the permits. These bonds have to be kept as a guarantee of the PV plant construction, and may be executed by the Caja General de Depósitos, if the depositor did not commission the PV plant within the 1 year granted, for having damaged third potential parties by booking this PV power in vain. Apparently, it did not affect the total 3,547 MW already connected to the grid, but did to the subsequent 2,500 MWn. We assess, conservatively, the energy used in association with these bonds (presumably as investments by the banks elsewhere) as 2,500 MW times 500,000 €/MWn = 1,250 million euros, deposited for 6 months to

**Table 6.15** Communications, control and management costs assumed for PV plants in Spain up to 2010

| Type of infrastructure | % of type of plants | Approximate plant size | | |
| --- | --- | --- | --- | --- |
| | | <2 MW 36% | 2–5 MW 20% | >5 MW 44% |
| | | No. of plants | No. of plants | No. of plants |
| | | 0.5 MWavg | 2.5 MWavg | 5 MW avg |
| Total | | 2,814 | 313 | 344 |
| Cost of comm, control and mgmt per plant/year | | 80% × 1,000 20% × 600 | 1,000 | 4,000 |
| Total economic costs in euros/year | | 2,588,880 | 313,000 | 1,376,000 |
| Total energy equiv. Costs in GWh/year | | 7.4 | | |

build a 1 MW typical PV plant. At a bank interest rate of between 1% and 2% per year, we use 1% of the bond value, depending on the purchasing power or the financial strength of the requesting entity (the more powerful, the lower the rate the financial costs will be). This would amount to 5,000 €/MWn per year. The economic cost may be calculated as 2,500 €/MWn. Therefore, the total cost of the bonds for the Spanish PV-installed plants in 2009 would be around 8 million euros.

There is also the necessity to ask the corresponding power utility in the region for the proper connection and access point of the given plant. These companies are asking about 2,000 €/MWn requested. This is supposed to help the electric distribution company to make the feasibility study for grid connection. Then, the total financial costs of these studies are 2,000 € times 4,043 MWn up to 2011 ≈ 8 million euros. Since this is less than 10 million euros for the whole PV life cycle, we ignore it ($a_{20} = 0$).

## *Other (Hardware)*

### $a_{21}$ Electrical Network/Power Lines Restructuring

A national electric network like the one in Spain consists of different sectors: generation and transport power lines (from 400 to 220 kV and lower, as per the new Spanish regulations); transformation (substations, transformers), distribution (power lines from 130 to 15 kV), consumption of low tension, and even commercialization efforts in the new liberalized market. These sectors are separate from the producer companies and entities we have been focusing on so far. It is difficult to assess the total value of the Spanish electric transport, transformation, and distribution, among other things, because the structure of the network has dramatically changed due to liberalization (privatization) of the market in the last few years and because the balances of the companies are transferred fast, changing assets from one company to another faster than we can grasp in this study. However, the latest data we could collect from public sources is given in Table 6.16.

The total declared assets owned by REE to the Comision Nacional del Mercado de Valores (CNMV) in 2009 were 6.2 billion euros, with 1.2 billion euros of net income; the amortization of fixed assets was 265 million euros and 330 million euros of net profit. As for the distribution companies, it exceeds the purpose of this book to enumerate in detail their changing assets and the new strategies of diluting their activities in different societies, after the liberalization processes or separating its interactions with foreign markets and other activities. We shall focus only on the distribution sectors of the main companies in Spain.

UNESA, the Spanish Association of the Electric Industry, composed of the main companies in the sector like Endesa, Iberdrola, Unión Fenosa, Hidrocantabrico ENEL-Viesgo, and REE stated in a presentation the following investments in the electric network of Spain (excluding REE) (Table 6.17).

**Table 6.16** Data of the transport situation and evolution of REE. Includes assets from other companies reported in 2004 and 2005

| Development of transmission grid | | | Development circuit bays at substations | | | Transformer capacity | | |
|---|---|---|---|---|---|---|---|---|
| Km circuit | 2004 | 2009 | Number busbar connections | 2004 | 2009 | Power (MVA) | 2004 | 2009 |
| 400 kV | 16,548 | 17,977 | 400 kV | 740 | 1,114 | Total | 37,216 | 66,259 |
| 220 kV and lower | 11,461 | 16,777 | 220 kV and lower | 1,188 | 2,271 | | | |
| Total | 28,009 | 34,754 | Total | 1,928 | 3,385 | | | |

**Table 6.17** Main investments in the Spanish electric grid in recent years

| In M€ | Generation | Transport and distribution | Renewables | Total UNESA |
|---|---|---|---|---|
| 2001 | 1,440 | 1,334 | 541 | 3,315 |
| 2002 | 1,821 | 1,576 | 406 | 3,803 |
| 2003 | 1,454 | 1,581 | 1,096 | 4,131 |
| 2004 | 1,472 | 1,823 | 1,148 | 4,443 |
| 2005 | 1,797 | 2,260 | 839 | 4,896 |
| 2006 | 2,410 | 2,346 | 910 | 5,666 |
| 2007 | 2,470 | 2,510 | 1,590 | 6,570 |
| 2008 | 2,570 | 2,610 | 1,770 | 6,950 |
| 2009 | 2,000 | 1,940 | 720 | 4,660 |
| Total | 17,434 | 17,980 | 9,020 | 44,434 |

*Source*: UNESA

A reasonable assumption is that the national transport and distribution network, including transformation and excluding generation and the renewable energies themselves, was worth some 100 billion euros. We assume that the duration of the electric network infrastructure is about 50 years. Therefore, the annual investment into the electric network would be a minimum of 2 billion euros. Multiplying that number by the 2.25% of the total electric demand that solar PV provided in 2009 equals 45 million euros/year, so that a conversion rate of a typical engineering cost of 3,960 MWh/Million euros, we derive $a_{20} = 178$ GWh/year, about 3.5% of the energy output. Should the solar PV penetration increase (in 2011 it reached 4%), the figure for this factor should be increased proportionally.

## $a_{22}$ Faulty Modules, Inverters, and Trackers

Spain had a painful, but very rapid learning curve associated with the huge flows of PV equipment in response to the very high subsidies and frantic efforts to meet PV installation dates. This PV equipment, mostly PV modules, inverters and trackers, was both produced locally and imported from "third world" countries such as China. One of the effects was a relatively large number of faulty modules and inverters. Spain has now, out of necessity, one of the most efficient and active number of specialized PV companies for checking the qualities of modules and their compliance with the required or announced specs. These companies maintain test labs, and certification centers, and they also specialized in inspections on-site, loss-adjuster's reports, arbitration in trials and legal disputes, technical and commercial assessments, etc. Generally speaking, these companies do not publish their data bases, as they are usually contracted by promoters and engineering, procurement and construction companies (EPCs), to check the quality of the PV equipment.

In a report of a specialized PV company, the company undertakes quality control of PV modules through visual inspection, peak power measurements, electrical isolation and thermographic inspections. Overall, it is hard to evaluate the quality of the modules, but this report provides an example of this analysis was made on 1% of the components (mostly PV modules, inverters and trackers) for two big plants of 19 and 13 MWp, respectively. The test results for this report detected defective modules from all manufacturers in both PV plants. Visual inspections showed high percentages (from 3.1% to 18.8%) of slight defects, which are not common in other mature industrial sectors. Peak power measurements found worrying results in both PV plants, with manufacturers with extremely high percentages of modules with peak power below tolerance (54.0% and 62.5% in Plant-A and 98% in Plant-B). Thermographic inspection showed worrying results as well, with average percentage of thermal defects of 18.7% in Plant-A and 8.7% in Plant-B. All the PV modules passed the electrical isolation test.

All of the solar PV energy produced in Spain in 2009 includes these (presumably) faulty or at least sub-par factors and modules. It is difficult to explore in full detail or to verify how many modules were returned under the guarantee because many contracts we have analyzed had such serious equipment flaws. This is likely to lower the energy production in the long term, which is another factor to consider. The official production figure that we use in this EROI analysis for the overall solar PV electricity production in Spain has already factored in the impacts of these failures. However, any module returned to the factory is a pure energy loss to the PV plant, even if the new module installed generates as per the specs. Thus, we include these losses in the denominator but not in the numerator of our EROI.

We estimate this factor as a return or rejection of 2% of the PV modules for the existing installed parks in Spain These modules may be partially repaired or totally scrapped. Assuming that the price of the PV modules installed in Spain up to 2009 is about 3 €/Wp, and considering that there were 3,670MWp installed, the total economic expense for faulty modules of the total turnkey projects of all installed PV plants in Spain up to 2009 would be about 3,670 MWp × 3€/Wp × 2% = 220 M€.

Inverters represent between 5% and 8% of the total cost of a PV turnkey project. They appear to have a similar percentage of failures due to mismatching with the modules imposed by constructors and promoters. This would represent some 23 million euros for faulty inverters of the total turnkey projects of all installed PV plants in Spain in 2009. Many of the failures are those of the IGBT's or solid state power devices. We have observed inverters completely burnt, including their housings. Summing all the above, we estimate the total losses for faulty modules and inverters as 243 million euros. This is about 10 million euros/year when averaged throughout the lifecycle of 25 years.

We do not consider here (it is still premature) other energy expenses of parts for the PV Plants that in some cases we know have suffered quality problems, such as oxidized nuts and bolts, or failed screws on modules or trackers to the structures, that had to be replaced or will have to be replaced in few years due to severe corrosion. In some cases, this would mean the physical labor of changing several million nuts and bolts in a single PV plant.

PV plants close to the sea may have the following problems:

1. "Tropicalization" of parts, where in tropical countries or in saline environments saline fogs envelope the PV plants many days in a year.
2. The need to manufacture them with the proper guarantees.
3. Therefore, they have an actual life span less than the theoretical standard of 25 years.

There have been frequent reports of modules from several suppliers, where the EVA encapsulant had a poor photo-stability, especially to the ultraviolet rays, causing migration of the material and loss of hermetism, which has resulted in the fast deterioration of the modules. It is too soon to assess these expenses, but in some cases the costs are significant.

We assume that 10 million euro/year so that at a conversion rate of a typical engineering cost of 3,960 MWh/Million euros, we derive are $a_{22} = 39.6$ GWh/year, about 0.75% of total energy output.

## $a_{23}$ Energy Costs Associated with the Network Stabilization Required with the Injection of Intermittent Loads

One of the largest problems with the expansion of renewable energies is related to the relation of intermittent loads relative to the main fossil-fueled systems. Energy analysts generally do not factor these problems into the life cycle analysis of the renewable energy systems, but they do have a certain impact. Fossil fuels are the principal fuels used by the utilities in part because they can be carefully titrated into the system to meet the fluctuating load demands. Any electricity corporation needs stability and security of supply under any usual circumstance. In contrast, renewable energies have an inherent risk of supply failure because they depend on natural cycles that, although can be predicted in the short term, cannot match the strict requirements of the electric demand. Some experts (e.g., Talbot 2012; Kunz and Balogh 2010) have studied this issue in depth recently and have tried to measure the negative impact on EROI of a source of energy, which is intermittent, considering that society has risk aversion and that intermittent energies represent an undoubted risk of supply. We will not enter into this analysis, but will consider the negative real life effects that the penetration of renewable energies into the Spanish grid has provoked in terms of the fossil fuels used to buffer the intermittencies of the PV cycle. The most important of these fuels is the combined cycle gas power plants, which reflect Spain's recent large use of natural gas, which in 2008 reached 34% of the total national electricity demand. The combined cycle gas power plants were the best able (after hydroelectricity) to come online fast to compensate for any undersupplies or go off-line fast for any oversupplies from renewable (PV or Wind) parks. But as the renewable systems were progressing and capturing a larger share of the national demand, and when both systems (gas and renewables) were combined, problems started to mount. The electricity market in Spain is one of the most liberalized (i.e., privatized) of Europe, had a lack of global holistic control of the energy production sources, and "left to market forces" the development of approaches to integrate the various energy sources, pushed by some premium tariffs for renewables or the threat of penalties for polluting emissions (e.g., $CO_2$), in the case of the investments in combined cycle gas-fired plants.

The business plans for combined cycle gas power plants are usually designed with the assumption to amortize them in their life cycle (25–40 years) due to their complexity, while requiring a smooth and efficient operation of a minimum 5,500 nominal hours in a year (Birol 2008). Although initially they reached some 4,000 nominal hours a year, suddenly things started to change, when renewables, led by wind, with a lot of intermittencies, passed 10% of the demand in 2008. Then the financial crisis came followed with the decline of the electric demand in Spain. Some forecasts from the power utilities and REE fearing that the 3,546 nominal hours of 2009, of the huge installed base of the combined cycle gas plants, could decrease to some 2,500 nominal hours/year, have already been converted into reality for 2010. Endesa and other combined cycle gas plants operators have shown a big alarm to the government. If this situation is sustained over time, it will make recovery of the investments in gas-fired plants by some of these companies more difficult.

Combined cycle gas plants have costs of about 1.2 euros per W installed. The plan in Spain has represented investments in the range of 20 billion euros. These investments, occurring at the start of their life cycle, are working less than half of the time as drafted in the business plans. This is in part because the combined cycle gas plants are now forced many hours in a year to work in "complementary intermittent regime" to compensate for the steep variations of wind and solar power: switching them off when there is oversupply if the wind blows and/or the sun shines and switching them on again if there is a calm wind and/or clouds in the skies. The head of the demand management in REE was quoted as saying that these gas plants are, in effect, manageable and can be modulated, but they have to be in standby mode and coupled to the grid several hours before a forecasted grid need. In fact, a night/day cycle, sometimes obliges these plants to switch off in the night and switched on in standby mode several hours before dawn when consumption ramps up. The natural gas used to keep up this "spinning reserve" is simply wasted, and it is obvious that these regimes of continuous and frequent switch on/offs shortens the expected life cycle of these complex dual gas cycle plants with complex thermal processes for energy secondary recovery considerably. Moreover, most of the natural gas consumed in Spain comes from Algeria, under long-term contracts of the "take or pay" type. Spain enjoyed very favorable prices per cubic meter of natural gas during that time, but now the growing idle times of gas-fired plants due to the increase in penetration of renewables has a cost in natural gas that has to be paid anyways, whether it is consumed or not.

It does not seem very sensible to appeal again to the fossil-fueled society, as some renewable energy associations have recently claimed (Expansion newspaper 2010). They have dared to ask the Ministry of Industry that most of the subsidies through premium tariffs, expected to amount to some 6.7 billion euros in 2010, be raised to 13.5 billion by 2020 to help them grow on a 7% annual basis throughout the period. They have asked the government to impose a combined tax of some 80 billion euros to be distributed on oil companies (60 billions) and gas companies (20 billions) to tax.

We assess the damage to the gas-combined cycle gas plants as 40% of the forecasted 20 billion euros business plan (8 billion euros), plus another coefficient of additional losses for frequent switch on/offs or interruptions, beyond the calculated original cycle to sustain the renewable energies intermittencies, valued at another 2 billion. Dividing this sum by the 25 years life cycle gives 400 million euros of damages per year to the combined cycle gas system, which we assume is 30% attributable to solar PV, for some 50 million euros/year so that at a conversion rate of a typical engineering cost of 3,960 MWh/Million euros, or two times the energy intensity of the national mean, we derive here ($a_{23} = 198$ GWh/year), about 3.9% of the energy output.

We have not included here the additional costs of the "take or pay" gas contracts, which are not being used, for lack of the originally foreseen number of hours. It is certainly an important issue, as Spain's main supplier is Algeria, and it goes beyond the scope of this book to discern, for instance, how much of the 268 Mtoes of imported natural gas of 2009 in Spain were subject to these type of contracts and at what prices they were signed or how much is the differential and how much can be attributable to the increase of renewables and subsequent increased idle time of the gas-fired plants. Besides, this factor is somehow mixed with $a_{10}$ (associated energy costs to injection of intermittent loads), and even this factor was more oriented to pump up and other storage facilities. But it is clear that the more $a_{10}$ costs, the less impact the $a_{23}$ factor will have today, the main one to stabilize the Spanish electric network versus the PV-installed park.

## $a_{24}$ Force Majeure, Acts of God and Others: Windstorms, Lighting, Flooding, Hailstorms

There are possible acts of nature that may harm the PV plants and affect both the generation and cost extra energy to replace the damaged equipment, besides the stealing and vandalism. The energy lost by these acts has already been analyzed and included in the ER part of this EROEI study. The energy to replace the damaged equipment is an extra energy cost that is ignored by the LCA or EPBT analysis.

We recognize the difficulty to analyze and assess at a national level the damages caused by nature. However, the Spanish media has shown completely flooded PV plants in the Guadiana river banks in the last heavy rainfalls of the winter of 2011. We have examined PV plants where the wind throughout time has loosened the screws and added increasing vibrations to the modules, probably provoking more module failures. There are PV plants also affected by lightning, lacking lightning rods. We have references to some HCPV plants that are installed in places where the wind exceeds the nominal upper limit speeds. This has forced HCPV plants to stop generating electricity during many more hours than originally planned. There are cases of HCPV plants that have been allowed to operate under wind speeds close to the upper limit, where the vibrations from the wind have significantly reduced the expected energy production. These incidents might only worsen as time passes.

We have no references of modules having been affected by hail, although Spain has some heavy hail storms each year. The tempered glass of modules are designed theoretically to resist hail of 1 cm diameter at some 20 m/second speeds, but sometimes the hail has the size between a golf and a tennis ball and the falling speed is a function of the weight. Just one big hail storm and the 25 years theoretical PV lifespan could evaporate. For lack of sufficient data, we will not give any energy input for this factor or energy addition to the "EI" denominator. We added this factor just to make our readers realize that there are factors never included that may need further and more detailed assessments in the future so that ($a_{24}$=0).

**Table 6.18** Summary of the energy investment (Ei or Ein) factors in a solar PV system as calculated for Spain (a new version will be sent)

| Factor | Typical energy invested additions in a PV system (EI or Ein) | MEuros/year | GWh equiv/year | Energy invested in ER × X |
|---|---|---|---|---|
| *Energy used on site* | | | | |
| $a_1$ | Accesses, foundations, canalizations and perimeter fences | | 56.6 | 0.011 |
| $a_2$ | Energy investments of evacuation lines and rights of way | 68 | 4.7 | 0.001 |
| $a_3$ | Operation and maintenance energy costs | 199 | 394 | 0.077 |
| $a_4$ | Module washing and/or cleaning | | 11.2 | 0.002 |
| $a_5$ | Self consumption in plants | | 28.2 | 0.005 |
| $a_6$ | Security and surveillance | 70 | 138.6 | 0.024 |
| *Energy used off site to manufacture ingots/wafers/cells/modules and some equipment* | | | | |
| $a_7$ | Module, inverters, trackers and metallic infrastructures (labor excluded) | | 608 | 0.120 |
| *Other energy expenses for on site and off site sine qua non activities for solar PV plants* | | | | |
| $a_8$ | Transportations. From local manufacturers to China | | 96 | 0.019 |
| $a_9$ | Premature phase out of unamortized manufacturing and other equipment | 37.5 | 148.4 | 0.028 |
| $a_{10}$ | Associated energy costs to injection of intermitent loads: pump up costs and/or other massive storage systems, if applied | 0 | 0 | 0.000 |
| $a_{11}$ | Insurances | 30 | 19.9 | 0.004 |
| $a_{12}$ | Fairs, exhibitions, promotions, conferences, etc. | 40 | 26.4 | 0.005 |
| $a_{13}$ | Administration expenses | 52 | 34.3 | 0.007 |
| $a_{14}$ | Municipality taxes, duties and levies (2–4% total project) | 21 | 14 | 0.003 |
| $a_{15}$ | Cost of land long term rent or ownership | 13.2 | 8.7 | 0.002 |
| $a_{16}$ | Circumstantial and indirect labor (not included in direct labor activities) | 11.2 | 22 | 0.004 |
| $a_{17}$ | Agent representative or market agent | 9.1 | 6 | 0.001 |
| $a_{18}$ | Equipment stealing and vandalism | 3 | 11.9 | 0.002 |
| $a_{19}$ | Communications, remote control and management | 4.3 | 17 | 0.0033 |
| $a_{20}$ | Pre-inscription, inscription, registration bonds and fees | 0 | 0 | 0.000 |

(continued)

**Table 6.18** (continued)

| Factor | Typical energy invested additions in a PV system (EI or Ein) | MEuros/year | GWh equiv/year | Energy invested in ER × X |
|---|---|---|---|---|
| $a_{21}$ | Electrical network/power lines restructuring | 45 | 178 | 0.035 |
| $a_{22}$ | Faulty modules, inverters, trackers | 10 | 39.6 | 0.0075 |
| $a_{23}$ | Associated energy costs to injection of intermitent loads: network stabilization associated costs (combined cycles) | 50 | 198 | 0.039 |
| $a_{24}$ | Force majeure acts of god and others: wind storms, lighting, storms, flooding, hail | 0 | 0 | 0.000 |
| $a_1$ to $a_{24}$ | Total invested energy factors | | 2,065.3 | 0.408 |

## Summary of Energy Costs

The sum of all energy used on yearly basis to generate Spain's PV electricity, calculated using our broad boundaries and with the assumption that wherever money is
spent energy must flow, is a total of 2,065.3 GWe of net useful energy to society
(Table 6.18). This equals 40.8% of all the yearly energy generated by the solar PV
plants of Spain.

# Chapter 7
# Results, Sensitivity Analysis, and Conclusions

## Energy Return on Investment for Spanish Photovoltaic Energy in 2008

Our best estimate of the energy return on investment for Spanish photovoltaics in 2009–2011 can now be derived by dividing the net electric energy equivalent output of 5,069 GWh (Table 5.1, 6024 GWh/year Gross energy as modified per the ratio in figure 5.1 Chap. 5), by the direct sum all necessary energy inputs (2,065.3 GWh, as shown in Chap. 6, table 6.18). The result is an EROI of 2.45 thermal units of energy for one thermal unit invested. We believe that this is close to the "standard EROI" as advocated by Murphy et al. (2011).

## *Sensitivity Analysis*

As should be obvious to any reader, there is considerable uncertainty and even ambiguity in this number, especially as related to the inputs that are used. Earlier analyses (e.g., Raugei et al. 2012) used only the energy cost of the modules and inverters and some other equipment and obtained rather large EROIs of up to 19–38 to 1. But as we have detailed in Chap. 5, there are many other inputs, and we have done our best to include them in our analysis. And, as we showed in Chap. 4, when we simply derived total energy costs by multiplying the total monetary cost of a one GW project by the energy intensity of the Spanish (or world) economy, we get an EROI of 2.41:1, about the same as what we calculated above.

Nevertheless, there remains a great deal of uncertainty in our estimates. We next address a number of these and calculate new EROIs accordingly so that the reader can choose what he or she wishes as the best EROI. Some of these operations would operate to decrease the EROI and some to increase it:

P.A. Prieto and C.A.S. Hall, *Spain's Photovoltaic Revolution: The Energy Return on Investment*, SpringerBriefs in Energy, DOI 10.1007/978-1-4419-9437-0_7, © Pedro A. Prieto and Charles A.S. Hall 2013

1. Remove special effects of the recent Spanish economic situation. A number of costs were a result of the rush into the PV business occasioned by the very high government subsidies of 2006–2008. We assume that this affected factors $a_8$ (transportation), $a_9$ (premature phase out), slightly $a_{10}$ (intermittent loads), $a_{12}$ (exhibits), $a_{13}$ (administration), $a_{14}$ (Municipality Taxes and others) and $a_{20}$ (registration fees) or $a_{14}$ (some taxes). We assume as a maximum value half of $a_8$ and $a_9$ and one quarter of the others. Adding all these together would be 48+74.2+0+6.6+8.6+3.5+0 =140.9 GWh/year instead of the previously assesed 319.1 GWh/year for these concepts. This is 263 GWh/year less. So correcting for the effects of the special Spanish conditions of 2008 would result in an increase in our calculated EROI from 2.45:1 to 5,069/(2,065.3–278.2)=2.84:1.

2. Correct for the higher energy quality of the output. Most of the inputs to the production of the PV systems are fossil fueled, and the output is high-quality electricity. If we assume that electricity is worth three times what fossil energy is (and assuming that it is used for high-quality functions such as lights and computers and not space heating), then we might conclude that the quality-corrected EROI is 7.35:1. But this is a double edged argument. It assumes that PV systems replace already existing electricity generated by fossil or nuclear fuels. The world consumes 59 EJ in electrical form, but a total of 509 EJ of primary energy[1]. If solar PV systems would have to replace all other non electrical activities, then the 'transformity' will operate in exactly the other way around with respect to quality and suitability for all of them that would require an energy carrier (i.e. merchant fleet, armies, aviation, mechanized agriculture, heavy machinery for mining or civil works, or a big portion of land transport), thus making the EROI going probably close to 1:1. But here we assume it is all used for high quality functions.

3. Removal of financial services. Financial and investment services might not be thought of as relating directly to costs of a project (but we think they are). Removing $a_{11}$, $a_{12}$, $a_{13}$, $a_{14}$, $a_{15}$, $a_{17}$, and $a_{20}$ would remove (19.9+26.4+34.3+1 4+8.7+6+0)=109.3 GWh from the denominator, again changing the EROI to about 2.46:1. On the other hand, we have not included the financial costs of the credits and leasing for the solar PV plants anywhere in our business or financial assessments. Nor have we mentioned the credits and financial operations for manufacturers of solar PV modules and associated equipment. We shall not give the details of our assessment, but given that about 75% of the total turnkey price of the plants and related costs were financed we did assess the energy equivalent on interest (not the principal). The amount is about 5 billion euro in the 10 years following the financial contracts. At 660 MWh/ Million euro this is an energy equivalent of 3,300 GWh. If this amount is added to the denominator, then the resulting EROI will be 5,069 GWh/(2,065 +3,300), lower than 1 and therefore, a sort of "net energy sink". We did not

---

[1]Ref. 2011 World Economic and Social Survey UN. Sankey Diagram ellaborated by Cullen and Allwood, 2009.

include this value due to the difficulty in discriminating partially embodied financial investments in the factors we analyzed and to avoid double counting. But certainly banks, bankers, and their employees and investors are consuming energy in our system.

4. Certainly, labor would not be available if laborers were not paid. And their paycheck would have no meaning if energy were not available to give meaning to the laborer's paycheck. Generally, we assume that this paycheck money uses about the national mean (1.99 kWh/euro) to make the goods and services represented by the paycheck. Probably some of the labor is already considered with respect to our means of assessment of, for example, financial services, but perhaps not since that was not included in, for example, the University of Illinois studies. But we certainly did not include the energy of labor in the energy used to build or run the machinery. A Spanish Association of Solar PV industries (AEF), released a report indicating that *"The photovoltaic sector is the most intensive among the clean energies in job creation: between 7 and 11 person per installed MW, versus 2.7 persons/MW in other sectors"*. Another solar PV industry association, ASIF (today UNEF) released a report in july 2009 indicating that there were 41,700 people employed in the sector, from them 15,400 full time and 26,300 part time. We use the smaller estimate of labor in the two reports and assume that part timers are working 1/3 of a full time worker. This represents about 24,166 full time equivalent workers in the sector at 86 MWh/active worker (national mean). If we included our total labor cost (with some double counting), this would add 2,078 GWh to the denominator, so the EROI ratio would be 5,096/(2,065+2,078) or 1.22:1.

## Discussion

The Spanish PV experience was either very successful or a fiasco depending upon how you wish to evaluate it. Certainly the original objectives of greatly expanding the input of PV power to the Spanish economy were met. PV and the wind power encouraged may protect the Spanish economy from future oil shocks. But the cost in euros and energy has been enormous and clearly has contributed to the current Spanish economic crises.

This situation could have been avoided at the start by undertaking a systems analysis of the program, bringing together financiers, energy analysts, and systems thinkers, before it was undertaken. A quantitative flow chart of the entire monetary and energy input process would have shown the possibility of the over subscription that ensued and the very large financial and energy costs for what is really a very limited addition to the electricity supply of Spain.

We are certain that many of our readers will be critical of the broad inclusiveness in our energy costs and of the methods by which we derive energy costs. We believe that the absence of any I–O analysis of the Spanish economy—let alone adding in energy analysis—makes it very difficult to come up with much better numbers. But we would be happy to see someone do a better job and will take no offense if our preliminary analyses are shown to be in substantial error, although we think this is quite unlikely.

Is the EROI enough to run modern society? It would seem that the most generous total assessment of the complete energy inputs needed for the Spanish PV situation in 2008 gives an EROI of no greater than about 7:1. Is that a high enough EROI to run a modern society? This was discussed in Chap. 1, but clearly the answer is "marginal at best." It is important to mention that another EROI analysis, undertaken by Raugei et al. (2012), have come up quoting much higher EROI's in the ranges of 5.9:1 for mono-c Si and multi-c Si, 9.4 for Ribbon Si, and 11.8:1 for CdTe. While we commend their efforts, we think that our EROI analysis result of 2.45:1 on Spain's PV revolution has been obtained using a more comprehensive EROI approach to analyzing all the factors that have an influence in making photovoltaic systems a reality for Spain.

## Conclusion

Thus, we are left with a somewhat ambiguous picture of solar PV energy in Spain. Certainly, we can conclude that PV systems can have serious energy and EROI constraints, as they would have significant material constraints if they were able to increase exponentially (Gupta 2012). Their EROI is high enough to allow the existence of human endeavors, but only so at a very limited scale, if not subsidized by fossil fuels. Here, solar PV has appeared more as a "fossil fuel extender" than an energy source able to self-breed its own sector while providing a significant net energy surplus to our modern global society. The low EROI and high fossil energy dependence of PV means that they are hardly carbon neutral and may release at least a third as much $CO_2$ as other systems per delivered unit of energy. As it stands all PV, wind power and nontraditional biomass energies contribute far less than 1 % of our global energy use, and given the continued expansion of traditional fossil fuels, it is clear that renewables are not displacing fossil fuels but just adding to the mix. Presumably people are using more fossil than PV energy because of the low EROI of the latter, which makes the energy and financial payback time very long. Thus, while we, like many others, believe that we must turn to solar energy eventually, this will be a difficult transition and likely one that will be very costly. We think it critical for advocates of solar power to examine ALL the energy inputs and design systems that are far less fossil fuel dependent than was the system we analyzed. Given the many things we have learned in Spain about the inefficiencies of the administrative programs, this might not be so difficult.

Finally the low EROI that we derived is not an argument to "kill" the expansion of photovoltaic systems. For one thing such detailed analyses have not been undertaken for other energy systems, whose EROI's generally are developed using a less comprehensive analysis than what we do here. So maybe the evaluation that we provide of an operational PV system's EROI should be evaluated against perhaps half the EROIs of conventional fossil fuels, which might be a ballpark estimate of the values derived if such comprehensive analyses were done for coal, oil and so on (see a start of this in Hall et al. 2009). Certainly there must be be a much more comprehensive, objective such analysis undertaken if we are ever to understand well what energy choices are before us.

This book is written assuming fossil fuels, not sunlight, as the fuel we need to invest. We do know that if we take three heat units of coal we can generate one unit of high value electricity in a thermal plant. If we were to instead invest three heat units of coal into a PV system in Spain it would yield some 7.35 units of high value electricity; that is 7.35 times more than from burning that coal in a thermal plant. The problem is that the first delivers the electricity immediately; the problem for the latter is that these units are delivered over 25 years and need an up front fossil fuel investment of about 2 thermal units in the first year for the solar PV system and the third thermal unit along the 25 years for O&M and other recurrent energy expenses.

The asymmetric dependence needs to be considered. One can imagine our present society without solar PV systems, but certainly not without oil, at least in the immediate or mid term time horizon, before oil abandons us. The first big energy takeover of coal from biomass, as we have shown in figure 1.2 was started at the beginning of the 20th Century and with a global consumption of less than 40 EJ. The second big energy transition, when oil passed coal, took place in the sixties of the last century, with a global consumption of about 140 EJ. In both cases, this transition took place when the earlier energy source was still growing and supporting the coming one. The challenge today is that the energy source we need for the future would have to have the intensity and versatility to make the transition at 510 EJ of global consumption, at the very time when the production of the main energy source, oil, is not only not growing but clearly entering a post peak situation. The long time to recover the energy investment is not a minor issue when the solar PV systems are seen as "renewable energies" rather than as "non renewable and finite systems, able to capture a portion of renewable energy flows"; that is, not a "free fuel" as some consider them.

The potential disruption provoked by the growing scarcity of fossil fuel energy resources may greatly erode the ability to generate solar PV systems, still heavily underpinned with fossil fuels. Several of the factors in the numerator could go to zero if social instability surges.

So we wonder if the market disincentive we observe for solar PV is not so much from its low EROI but rather has to do with the discount rate, the time value of money. It also comes down to the related issue of whether financially-stressed people will give up money (or energy) now at the rates mentioned for theoretical, but not necessarily granted greater benefits for their children. These are very difficult issues far beyond our expertise. But they will be amongst the most important issues society will face as our principal main high-quality fuel resources are increasingly exhausted.

On the other hand our analysis shows that there is a great deal of room for decreasing energy costs. Addressing this plus continual advances in module efficiency could give solar PV a considerable role in the future. Whether this can be undertaken in the current political-economic environment is another issue.

# References

ASPO (Association for the Study of Peak Oil). Foundation Charter.

Balogh, S. and H. Kunz Applying Time to Energy analysis. The oil drum http://www.theoildrum.com/node/7147

Birol, F. 2008. *World energy outlook*. Paris: International Energy Agency. http://www.iea.org/textbase/nppdf/free/2008/weo2008.pdf. Accessed 19 Sep 2012.

Bullard, C., and B. Hannon. 1973. The energy costs of goods and services. *Energy Policy* 3: 263–278.

Bullard, C.W., B. Hannon, and R.A. Herendeen. 1975. *Energy flow through the US economy*. Urbana: University of Illinois Press.

Campbell, C., and J. Laherrere. 1998. The end of cheap oil. *Scientific American*. 278: 78–83.

Canadian Solar. 2012. http://www.canadiansolar.com/en/products/product-documentation/product-documentation.html. Accessed 19 Sep 2012. See warranty zip or pdf.

Carnegie Mellon Energy Use Calculator. 2012. http://www.eiolca.net/cgi-bin/dft/use.pl?searchTerms=&newmatrix=US430CIDOC2002. Accessed 19 Sep 2012.

Central Intelligence Agency (CIA). 2012. *The world factbook: Spain*. Updated 10 September 2012. https://www.cia.gov/library/publications/the-world-factbook/geos/sp.html. Accessed 15 Sep 2012.

Cleveland, C.J. 2005. Net energy from the extraction of oil and gas in the United States. *Energy: The International Journal* 30(5): 769–782.

Cleveland, C.J., R. Costanza, C.A.S. Hall, and R. Kaufmann. 1984. Energy and the United States economy: a biophysical perspective. Science 225: 890–897.

Comision National de Energia (CNE). 2012. *Información Estadística sobre las Ventas de Energía del Régimen Especial* (in Spanish). http://www.cne.es/cne/Publicaciones?id_nodo=143&accion=1&soloUltimo=si&sIdCat=10&keyword=&auditoria=F. Accessed 16 Sep 2012.

Ministerio de Industria, Turismo y Comercio. Registro de preasignación de retribución para instalaciones fotovoltaicas. Segundo trimestre de 2010.

Expansion Newspaper. 23 June 2010. *Las renovables quieren que las petroleras les paguen las subvenciones*. http://www.expansion.com/2010/06/23/empresas/energia/1277244633.html. Accessed 19 Sep 2012.

Fthenakis, V., and E. Alsema. 2006. Photovoltaics energy payback times, greenhouse gas emissions and external costs: 2004–early 2005 status. *Progress in Photovoltaics: Research and Applications* 14: 275–280. doi:10.1002/pip. 706.

Fthenakis, V.M., and H.C. Kim. 2011. Photovoltaics: life-cycle analyses. *Solar Energy* 85(8): 1609.

Fthenakis, V.H.C. Kim, R. Frischknecht, M. Raugei, P. Sinha, and M. Stucki. 2011. Life cycle inventories and life cycle assessment of photovoltaic systems. International Energy US Energy Investment Agency (IEA) PVPS Task 12, Report T12-02:2011. http://www.bnl.gov/pv/files/pdf/226_Task12_LifeCycle_Inventories.pdf. Accessed 19 Sep 2012.

Gadir Solar (Oerlikon Technology). 2012. http://www.gadirsolar.es/en/documentos/FichaSIRENGfeb2011.pdf. Accessed 19 Sep 2012.

Gagnon, N., C.A.S. Hall, and L. Brinker. 2009. A preliminary investigation of energy return on energy investment for global oil and gas production. *Energies* 2(3): 490–503.

Grandall, L., C.A.S., Hall, and M. Hook. 2011. Energy return on investment for Norwegian oil and gas in 1991–2008: Sustainability: Special Issue on EROI. Pages 2050–2070.

Guilford, Megan C., Charles A.S. Hall, Peter O'Connor, and Cutler J. Cleveland. 2011. A new long term assessment of Energy Return on Investment (EROI) for U.S. oil and gas discovery and production. *Sustainability* 3(10): 1866–1887.

Gupta, A.K., and C. Hall. 2011. Energy Costs of Materials Associated with the Exponential Growth of Thin-Film Photovoltaic Systems, in Fundamentals of Materials for Energy and Environmental Sustainability, edited by D.S. Ginley and D. Cahen (Materials Research Society, Warrendale, PA, and Cambridge Univ. Press, Cambridge, England), 2011.

Hall, C.A.S. 1972. Migration & metabolism in a temperate stream ecosystem. *Ecology* 53(4): 585–604.

Hall, C.A.S., M. Lavine and J. Sloane. 1979. Efficiency of energy delivery systems: Part I. An economic and energy analysis. Environ. Mgment. 3 (6): 493–504.

Hall, Charles A.S. 2011. Introduction to special issue on new studies in EROI (Energy Return on Investment). *Sustainability* 3(10): 1773–1777.

Hall, C.A.S., and K. Klitgaard. 2011. *Energy and the wealth of nations: understanding the biophysical economy*. New York: Springer.

Hall, C.A.S., C. Cleveland, and M. Berger. 1981. Energy return on investment for the United States Petroleum, Coal and Uranium. In *Energy and ecological modeling, Symp. Proc.*, ed. W. Mitsch, 715–724. Amsterdam: Elsevier Publishing Co.

Hall, C.A.S., C.J. Cleveland and R. Kaufmann. 1986. Energy and Resource Quality: The ecology of the economic process. Wiley Interscience, NY.

Hall, C.A.S., S. Balogh, and D.J.R. Murphy. 2009. What is the minimum EROI that a sustainable society must have? *Energies* 2: 25–47.

Hamilton, J. 2009. Causes and consequences of the oil shock of 2007–08. In *Brookings papers on economic activity*, ed. D. Romer, and J. Wolfers, 1–68. Conference series. Washington, DC: Brookings Institute.

Hispanidad. 2010. *Las subvenciones a las renovables destrozan el ciclo combinado*. http://www.hispanidad.com/noticia.aspx?ID=133710. Accessed 19 Sep 2012.

Hubbert, M.K. 1969. *Energy resources*. In *Resources and man*, ed. W.H. Freeman, 157–242. San Francisco: National Academy of Sciences.

Instituto para la Diversidad y Ahorro de Energia (IDAE). 2011. *Plan De Energia Renovables 2011–2020*, 147 (Table 10.a). http://www.idae.es/index.php/mod.documentos/mem.descarga?file=/documentos_11227_PER_2011-2020_def_93c624ab.pdf. Accessed 19 Sep 2012.

Jacobson, Mark Z., and Mark A. Delucchi. 2011. Providing all global energy with wind, water, and solar power. Part I. Technologies, energy resources, quantities and areas of infrastructure, and materials. *Energy Policy* 39(3): 1154–1169.

Kunz, S. and H. Kunz. The Fake fire Brigade Part IV. The Oil Drum http://www.theoildrum.com/node/6957 Tverberg Gail.

Little, D. Arthur. 2001. *Potential reduction costs in PV systems*. Cambridge: Cambridge Consultants Limited. http://webarchive.nationalarchives.gov.uk/tna/+/http://www.dti.gov.uk/renewables/publications/pdfs/sp200320.pdf/. Accessed 19 Sep 2012.

Miliarium. 2008. *Plan Hidrologico Nacional*. http://www.miliarium.com/Bibliografia/Monografias/PHN/Situacion_Espana.asp. Accessed 19 Sep 2012.

Ministry of Environment. 2009. *Perfil Ambiental de España 2009*, 67 (Chapter 2.2). Agua: Ministry of Environment. http://hispagua.cedex.es/documentacion/documentos/Perfil_Ambiental_Espana_2009/Agua.pdf. Accessed 19 Sep 2012.

Ministry of Environment. 2009. *Perfil Ambiental de España 2009*, 67 (Chapter 2.2). Agua: Ministry of Environment. http://hispagua.cedex.es/documentacion/documentos/Perfil_Ambiental_Espana_2009/Agua.pdf. Accessed 16 Sep 2012.

Murphy, D.J., and C.A.S. Hall. 2011. Energy return on investment, peak oil, and the end of economic growth. *Annals of the New York Academy of Sciences. Special Issue on Ecological Economics* 1219: 52–72.

Murphy, David J., Charles A.S. Hall, Michael Dale, and Cutler Cleveland. 2011. Order from chaos: a preliminary protocol for determining the EROI of fuels. *Sustainability* 3(10): 1888–1907.

Odum, H.T. 1973. *Environment, power and society*. New York: Wiley Interscience.

Organisation Internationale des Constructeursd' Automobile (OICA). 2012. *World motor vehicle production ranking by country 2010–2011*. http://oica.net/wp-content/uploads/total-2011-august-2012.pdf. Accessed 15 Sep 2012.

PANER, Plan de Acción Nacional de Energías Renovables de España. 2011. *Official draft*, 147 and 30.

Raugei, M., P. Fullana-i-Palmer, V. Fthenakis. 2012. The energy return on energy investment (EROI) of photovoltaics: Methodology and comparisons with fossil fuel life cycles Energy Policy 45: 576–582.

Red Electrica De Espana (REE). 2012. http://www.ree.es/ingles/home.asp. Accessed 19 Sep 2012.

Simmons, Matthew. 2007. PEAK OIL: is it real? When might it occur? In *Kayne Anderson energy funds partners' meeting*, Houston, 12 February 2008 (Chart 35).

Smil, V. 1994. *Energy in world history*. Boulder: Westview Press.

Smil, Vaclav. 2006. Energy at the crossroads. Background notes for a presentation to the Global Science Forum Conference on Scientific Challenges for Energy Research, Paris, May 17 and 18.

Suntech. 2012. http://eu.suntech-power.com/images/stories/pdf/warranty_feb_2012/2012%20standard%20warranty%20europe%20version%20v2%20en.pdf. Accessed 19 Sep 2012. See paragraph 2. Performance warranty.

Tainter, J. 1988. *The collapse of complex societies*. Cambridge: Cambridge University Press.

Talbot, David 2012. The Great German Energy Experiment. Technology Review (MIT). June 18th 2012.

Vector Cuatro. 2012. http://www.vectorcuatro.es/index.php/es/noticias/8-noticias/19-vector-cuatro-renegocia-las-perdidas-de-mt-de-sus-plantas-gestionadas-bajandolas-mas-de-un-50. Accessed 19 Sep 2012.

White, L.A. 1949. *The science of culture*. New York: Farrar, Straus, and Co.

# Index

P.A. Prieto and C.A.S. Hall, *Spain's Photovoltaic Revolution: The Energy Return on Investment*, SpringerBriefs in Energy, DOI 10.1007/978-1-4419-9437-0,
© Pedro A. Prieto and Charles A.S. Hall 2013